Marine Ecological Field Methods

Marine Ecological Field Methods

A Guide for Marine Biologists and
Fisheries Scientists

Edited by

Anne Gro Vea Salvanes
Department of Biology, University of Bergen
Bergen, Norway

Jennifer Devine
Institute of Marine Research (IMR)
Bergen, Norway

Knut Helge Jensen
Department of Biology, University of Bergen
Bergen, Norway

Jon Thomassen Hestetun
Department of Biology, University of Bergen
Bergen, Norway

Kjersti Sjøtun
Department of Biology, University of Bergen
Bergen, Norway

Henrik Glenner
Department of Biology, University of Bergen
Bergen, Norway

Registered Office(s)
John Wiley & Sons, Inc., 111 River Street, Hoboken, NJ 07030, USA
John Wiley & Sons Ltd, The Atrium, Southern Gate, Chichester, West Sussex, PO19 8SQ, UK

Editorial Office
The Atrium, Southern Gate, Chichester, West Sussex, PO19 8SQ, UK

For details of our global editorial offices, customer services, and more information about Wiley products visit us at www.wiley.com.

Wiley also publishes its books in a variety of electronic formats and by print-on-demand. Some content that appears in standard print versions of this book may not be available in other formats.

Library of Congress Cataloging-in-Publication data applied for

ISBN: 9781119184300

Cover Design: Wiley
Cover Images: Northeast Atlantic mackerel swimming, Courtesy of Leif Nøttestad;
Intertidal community at awave-exposed site, Courtesy of Kjersti Sjøtun

Set in 10/12pt Warnock by SPi Global, Pondicherry, India

Contents

3   Sampling Gears and Equipment *75*
Anne Gro Vea Salvanes*, Henrik Glenner*, Dag L. Aksnes, Lars Asplin, Martin Dahl, Jennifer Devine, Arill Engås, Svein Rune Erga, Tone Falkenhaug, Keno Ferter, Jon Thomassen Hestetun, Knut Helge Jensen, Egil Ona, Shale Rosen and Kjersti Sjøtun
* Lead authors; co-authors in alphabetical order

List of Contributors

Dag L. Aksnes
Department of Biology,
University of Bergen (BIO, UiB),
Bergen, Norway

Lars Asplin
Institute of Marine Research (IMR),
Bergen, Norway

Martin Dahl
Institute of Marine Research (IMR),
Bergen, Norway

Jennifer Devine
Department of Biology,
University of Bergen (BIO, UiB),
Bergen, Norway;
Institute of Marine Research (IMR),
Bergen, Norway

Arill Engås
Institute of Marine Research (IMR),
Bergen, Norway

Tone Falkenhaug
Institute of Marine Research,
Flødevigen, His,
Norway

Svein Rune Erga
Department of Biology,
University of Bergen (BIO, UiB),
Bergen, Norway

Keno Ferter
Institute of Marine Research (IMR),
Bergen, Norway

Henrik Glenner
Department of Biology,
University of Bergen (BIO, UiB),
Bergen, Norway

Jon Thomassen Hestetun
Department of Biology,
University of Bergen (BIO, UiB),
Bergen, Norway

Mette Hordnes
Department of Biology,
University of Bergen (BIO, UiB),
Bergen, Norway

Ragnhild Aakre Jakobsen
Hunstadsvingen,
Bergen, Norway

Knut Helge Jensen
Department of Biology,
University of Bergen (BIO, UiB),
Bergen, Norway

Frank Midtøy
Department of Biology,
University of Bergen (BIO, UiB),
Bergen, Norway

Leif Nøttestad
Institute of Marine Research (IMR),
Bergen, Norway

Egil Ona
Institute of Marine Research (IMR),
Bergen, Norway

Michael Pennington
Institute of Marine Research (IMR),
Bergen, Norway

David John Rees
Department of Biology,
University of Bergen (BIO, UiB),
Bergen, Norway

Shale Rosen
Institute of Marine Research (IMR),
Bergen, Norway

Anne Gro Vea Salvanes
Department of Biology,
University of Bergen (BIO, UiB),
Bergen, Norway

Kjersti Sjøtun
Department of Biology,
University of Bergen (BIO, UiB),
Bergen, Norway

Arved Staby
Institute of Marine Research (IMR),
Bergen, Norway

Foreword

Despite covering over 70% of the surface of the planet, the marine environment is less accessible, and thus less well-known than terrestrial habitats. A variety of technologies allow for marine field studies on environments ranging from the shallow nearshore to depths of thousands of meters, on individuals, populations, communities, and ecosystems. This book describes marine ecological sampling equipment, methods, and analysis, ranging from physical parameters to fish, microalgae, zooplankton, benthos, and macroalgae. It will be useful for graduate students and early-stage professionals in marine biology and fisheries, even those not directly involved in fieldwork, by giving an overview of marine biological data collection, handling and analysis.

This handbook provides a guide to the use of marine ecological sampling methods used for pure research and for fisheries management purposes. The book covers survey and sampling design, sample and data collection and processing, and data analysis. The research question and characteristics of the organisms and habitat dictate what sampling equipment is required. Information is included on sampling equipment, ranging from those that are useful in shallow nearshore areas, such as bottles, secchi discs, and gillnets or beach seines to those deployed from large research ships for studies offshore, such as remotely operated vehicles (ROVs), fishing trawls, and hydroacoustics, or remote observation using satellites.

The development of this book started at the Department of Biology at the University of Bergen in 2011; when due to lack of suitable literature, students attending a marine field course were provided with short handouts. The handouts became more and more advanced from year to year. In 2014 the publisher Wiley became aware of the initiative and invited us to write a textbook for broader use. The six editors of the book have, over several years, been involved in the writing and development of the book project. As we came across additional themes relevant for the handbook, and that we ourselves felt we did not know well enough, we invited experts from our network at the Institute of Marine Research and the Department of Biology at the University of Bergen to contribute as co-authors. The editors have produced text, and in addition taken the lead on the structure, contents, and in the editing of the entire manuscript. All editors have worked on the full text. A.G.V. Salvanes has had the main responsibility for coordinating the work, J. Devine has had the final edit on all chapters, J.T. Hestetun was mainly responsible for keeping references organized and for quality evaluation of figures.

All artwork was produced by R. Jakobsen. We hope the handbook will help reader to plan and execute fieldwork to answer research questions, and provide basic knowledge of the most common methods for collecting field data for modern marine research. We also hope the handbook will enable readers to explain and evaluate the principles of different sampling approaches, their strengths and weaknesses, and not least how to process, catalog, and interpret collected field samples and experimental data.

The provided R code with this book (http://filer.uib.no/mnfa/mefm/) is free software; you can redistribute it and/or modify it under the terms of the GNU General Public License as published by the Free Software Foundation; either version 3 of the License, or (at your option) any later version. If the code and data are used for teaching (or other) purposes, we ask those using the material to reference the textbook. The code is distributed in the hope that it will be useful, but WITHOUT ANY WARRANTY; without even the implied warranty of MERCHANTABILITY or FITNESS FOR A PARTICULAR PURPOSE. See the GNU General Public License for more details.

Bergen, January 20, 2017

Anne Gro Vea Salvanes
Jennifer Devine
Knut Helge Jensen
Jon Thomassen Hestetun
Kjersti Sjøtun
Henrik Glenner

Acknowledgements

The editors acknowledge the Olav Thon Foundation for funding which made the completion of this book possible. Thanks especially to the many students on our marine field courses that over years have inspired us to write this book. We are particularly grateful to the 2016 master students on the Ocean Science Course (BIO325) at the Department of Biology, University of Bergen. They tested out and gave us valuable input to improve the draft version of the book: M.V. Bjordal, A. Delaval, C. Djønne, N.E. Frogg, K.F. Furseth, S. Hjelle, J.S. Høie, I. Nilsen, D. Notvik, H. Seal, M.R. Solås, E. Tessin, and S. Tonheim. G.J. Macaulay, Institute of Marine Research and L.H. Pettersson; Nansen Environmental and Remote Sensing Center are thanked for professional help and production of topographic and remote sensing maps. We thank T. Klevjer, J.H. Vølstad, and K. Korsbrekke, Institute of Marine Research for comments and B.H. Bjørnhaug, Bergen Technology Transfer Office for help with contract issues. Many colleagues and companies are thanked for illustrations; Aanderaa, G. Anderson, Santa Barbara, Fagbokforlaget, Institute of Marine Research, T. Hovland, G. Macaulay, K. Mæstad, R.D.M. Nash, Ø. Paulsen, Scantrol/Deep Vision, H. Saivolainen, H.R. Skjoldal, Son Tec, E. Svendsen, T. Sørlie, G. Sætra, University of Bergen Library. We thank: A. Hobæk, Norwegian Institute for Water Research; C. Todt, Rådgivende Biologer AS; M. Malaquias, University Museum of Bergen; L. and P. Buhl-Mortensen, Institute of Marine Research, H.T. Rapp and K. Meland, University of Bergen, T. Dahlgren, University of Gothenburg, and U. Båmstedt, Umeå University, for valuable contributions to benthic studies. Our special thanks goes to the crew onboard the research vessels; RV Håkon Mosby, RV G.O. Sars, RV Hans Brattstrøm and RV Dr. Fritjof Nansen; the crew are experts and have deep knowledge on operating advanced as well as the simple gears used to sample marine organisms. We could not have done our research or field courses without their skills and support.

Acknowledgements

The editors acknowledge the Olav Thon Foundation for funding which made the completion of this book possible. Thanks especially to the many students on our marine field courses that over years have inspired us to write this book. We are particularly grateful to the 2015 master students on the Ocean Science Course (BIO325) at the Department of Biology, University of Bergen. They tested out and gave us valuable input to improve the draft version of the book. M.V. Bjordal, A. Delavé, C. Djønne, N.E. Frogg, K.E. Fuszah, S. Hjelle, J.S. Hole, I. Nilsen, D. Notvik, H. Seal, M.R. Solås, E. Teisin, and S. Tonheim, G.I. Macaulay Institute of Marine Research and L.H. Pettersson Nansen Environmental and Remote Sensing Center are thanked for professional help and production of topographic and remote sensing maps. We thank T. Kleiven, J.H. Volstad, and K. Enoksdale, Institute of Marine Research for comments and B.H. Bjørnstad, Bergen Technology Transfer Office for help with contract issues. Many colleagues and companies are thanked for illustrations, Andersa, G. Anderson, Santa Barbara, Fagbokhaget, Institute of Marine Research, T. Hovland, G. Macaulay, K. Mæstad, R.D.M. Nash, Ø. Paulsen, ScanturDeep Vision, H. Salvoleinen, H.R. Skofdal, Son Tec, E. Svendsen, T. Svendsen, G. Sætre, University of Bergen Library. We thank A. Hobæk, Norwegian Institute for Water Research, C. Todt, Rådgivende Biologer AS, M. Malaquias, University Museum of Bergen I. and P. Buhl-Mortensen, Institute of Marine Research, H.T. Rapp and K. Meland, University of Bergen, ... Dahlgren, University of Gothenburg, and U. Båmstedt, Umeå University, are valuable contributions to existing studies. For special thanks to the research oriented experiments used ... Heine Meldal KV CD ... R/V Hans Brattström and R/V Fangst, ... lectures on the ... and experimental new developments, the operational aspects of ... at the surface generation to single marine organisms. We could not have done our research and field courses without their skills and support.

1

The Marine Environment

Jon Thomassen Hestetun*, Kjersti Sjøtun*, Dag L. Aksnes, Lars Asplin, Jennifer Devine, Tone Falkenhaug, Henrik Glenner, Knut Helge Jensen and Anne Gro Vea Salvanes

The marine environment covers over 70% of the surface of the Earth, yet represents special challenges when it comes to scientific inquiry. When compared to terrestrial systems, the marine environment is much less easily accessible and, despite great effort, remains less well known. With the rise of the modern natural sciences, tools and methods have been continually developed to explore marine environments, from the littoral zone and nearshore environment to open waters and the shelf and abyssal seafloor. From tried and true collection equipment, often identical to or based on fishing gear, to new innovations in remotely controlled and autonomous vehicles, exploration of the underwater world is heavily dependent on the tools used.

Technological advancement now allows marine field studies to be conducted at all levels: from individuals to populations, to groups of populations, and to entire ecosystems. Habitats from the shallow nearshore to depths of thousands of meters are increasingly accessible; studies of interactions between specific organisms and physical and biological components are possible. The equipment used for sampling is dependent on the research questions asked and the characteristics and depth of the studied organisms and their habitat. Gears range from simple tools that are useful in shallow nearshore areas, such as bottles, secchi discs, and gillnets or beach seines to advanced equipment, such as remotely operated vehicles (ROVs), fishing trawls, and hydroacoustics deployed from large research ships for studies offshore and at greater depths. Even remote observation from space can be performed using satellites.

A characteristic transect from a continental landmass to the deep ocean includes nearshore environments that, depending on local geology, may consist of sandy beaches, cliffs or fjord systems. The continental shelf may stretch out some distance from the continental landmasses, gradually giving way to the continental slope, which descends down to the abyssal plains of the world's major

---

* Lead author; co-authors in alphabetical order.

Marine Ecological Field Methods: A Guide for Marine Biologists and Fisheries Scientists,
First Edition. Edited by Anne Gro Vea Salvanes, Jennifer Devine, Knut Helge Jensen,
Jon Thomassen Hestetun, Kjersti Sjøtun and Henrik Glenner.
© 2018 John Wiley & Sons Ltd. Published 2018 by John Wiley & Sons Ltd.

oceans. As an example, the western coast of Norway contains an elaborate fjord system with numerous deep basins divided by shallower sills, giving way to the Norwegian Channel and then the shallower continental shelf. To the southwest, the North Sea is a shallow sea on top of a continental shelf only, while to the northwest, the Norwegian Sea descends into a deep-sea basin which also contains the Mid-Arctic Ridge, separating the Eurasian and North American continental plates. Banks, seamounts and submarine canyons are features that add to the topographical complexity of this general system (Figure 1.1).

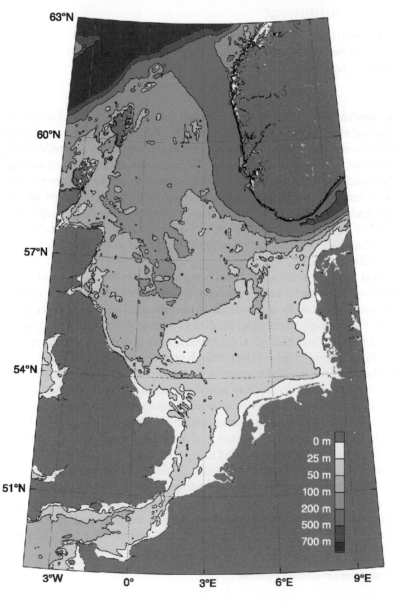

Figure 1.1 Topographic chart of the North Sea. *Source:* G. Macaulay, Institute of Marine Research, Norway.

Species composition changes with depth and distance from the coast, both for pelagic species and for organisms associated with the seafloor. Organisms are morphologically, physiologically, and behaviorally adapted to their environment through natural selection. Individuals with favorable genetic traits have increased breeding success than those lacking these traits (genetic adaptation). Some species are able to shift between environments and habitats, for instance benthic species with a pelagic egg and larval phase, or species that shift diurnally between different water depths (diel vertical migration, DVM). Diel vertical migration typically occurs between water masses with different properties in terms of light, temperature, oxygen, and salinity, requiring a physiological response from the organism. In general, effects of abiotic and biotic factors influence morphology, physiology, and behavior and thus how animals adapt to their habitat.

Examples of **abiotic factors** are the optical properties of the water column and include: light and the amount of suspended particles, which are important for visual predation; temperature, which regulates physiology, metabolic processes, and swimming activity; salinity, which affects physiology and osmoregulation; oxygen levels, which regulate respiration and metabolism and can limit reproduction or growth at low levels; and depth, which regulates pressure and affects buoyancy of fish that use swim bladders to obtain neutral buoyancy. Stratification of water masses, which often is seasonally dependent, limits nutrient availability in upper strata (the photic zone, as well as oxygen concentration in the lower strata or in isolated basins. Eutrophication and closeness to urbanized regions will also affect the level of primary production and the depth range where visual feeding is possible.

Biotic factors influencing the structure of marine communities and ecosystems include prey availability, predation, competition, and parasitism, and are regulated by direct or indirect access to production from lower trophic levels. Trophic communities in shallow waters benefit from readily available photosynthetic primary production, however, such production may be limited by nutrient availability. Organisms in deeper layers usually depend on energy and biomass from above either through migrating animals, transporting nutrients from surface waters to depths, or through the downward transport of debris, dead organisms, and particulate organic matter (POM). Because lower systems are dependent upon the upper regions, total biomass often decreases with depth. Population and individual growth potential will be further regulated through food access and competition. Access to reproduction (mates and spawning grounds) and reproductive behavior (nest spawning, demersal spawning, or pelagic spawning) will affect recruitment to populations. Presence of suitable nursery environments (e.g., coral reefs and kelp zone habitats) regulates survival of early life stages (larval stages of benthos and juvenile fish). Mortality risk (predator density, visibility, and size) in the habitats also changes with depth and distance from the coast.

Chapter 1 begins with a brief description of zonation in the pelagic and benthic realms, followed by a description of the topographies of coastal and fjord biotopes, the continental shelf and slope, and the deep ocean. These biotopes shape the habitats for bottom associated marine organisms. This is followed by a description of the physical characteristics of the pelagic ecosystem, including circulation of water masses in fjord ecosystems and a description of the light environment in marine waters. The chapter ends with an overview of temperate organisms (benthos and fish) that inhabit the littoral, sublittoral, continental shelf and slope, deep fjords, and the deep sea.

1.1 Marine Habitats

1.1.1 The Pelagic and Benthic Realms

The oceans are commonly divided into the pelagic and benthic realms. The pelagic realm refers to the body of water from above the seabed to the surface of the water. The organisms swimming or floating in this water column are termed pelagic and can be roughly divided into nekton, able to control their position in the water masses, and plankton. Traditionally the pelagic realm is subdivided into five zones:

1) The epipelagic: from the surface to about 200 meters and where the amount of UV light from the sun still allows photosynthesis.
2) The mesopelagic zone: from about 200 to approximately 1000 meters. The twilight zone where the organism still might be able to detect sunlight, but at which depth the energy from UV light is too limited for photosynthesis.
3) The bathypelagic zone: from 1000 to about 4000 meters, where no sunlight remains.
4) The abyssopelagic zone: from about 4000 to 6000 meters. The average depth of the largest oceans in the world is largely contained in this zone – between the 3300 meters of the Atlantic Ocean to the 4300 meter average depth in the Pacific.
5) The hadopelagic zones: between 5000 to 6000 meters. These zones are found in relatively restricted areas like deep trenches to the deepest trench, the Mariana trench, which is about 11000 meters deep.

The benthic realm is defined as the bottom sediment or seabed of the ocean and the organisms in or on it are defined as the benthos. Organisms living in the benthic realm are living in close a relationship with the sediment, often permanently attached to it (epibenthos) or burrowing in it (endobenthos), while others, although they can swim, are never found far away from the seafloor, on which they are totally dependent (hyperbenthos or, in the case of fishes, demersal).

About 80% of the ocean floor consists of soft sediment, which can be designated as marine sediments of particle size ranging from mud to coarse sand (0.05 mm to about 1 mm in diameter). This entails that this soft-bottom substrate type defines the vast majority of habitats, from the high subtidal zone to the deepest part of the abyssal zone. Obviously, the term is very broadly defined, and soft sediments can be divided into a number of subhabitats, which are dependent on latitude, temperature and other local environmental factors, including a wide size range and a high diversity of associated organisms.

Ocean hard bottom areas, while less extensive, represent important habitats distinguished by differences in biota composition and dominating life strategies compared to soft-bottom counterparts. Hard bottom seafloor is often associated with specific topographical features such as for instance submarine canyons, seamounts or mid-ocean ridges, or other areas with stronger currents. It can provide substrate for large number of immobile organisms, and current activity can form the basis of filter-feeding communities.

The benthic realm is zoned by depth in a way that generally corresponds to the zones in the pelagic realm:

1) The intertidal zone: where land meets the sea. This has no parallel in the pelagic realm.
2) The sublittoral zone: defined as the area of the coast that, even at lowest tide, is always submersed to the extent of the continental shelves. The continental shelves extend to approximate depths of 200 meters. This corresponds to the epipelagic zone.
3) The bathyal zone: extends from 200 meters to approximately 4000 meters. This also includes the continental slope and corresponds to the mesopelagic and the bathypelagic zone.
4) The abyssal and hadal zones includes most of the ocean seafloor from 4000 meters to the deepest trench at 11 000 meters. These zones correspond, respectively, to the abyssopelagic and the hadopelagic zones in the pelagic realm.

1.2 The Coastal and Fjord Biotopes

Fjord systems are found in many areas of the world including Alaska, Chile, Greenland, Norway, and New Zealand. They have a complex topography that can include numerous narrow passages, and are often divided into basins divided by shallower sills, which are shallow ridges situated at the mouth of the fjord and are normally old moraines. The outer part of the coast consists of a number of islands and thus the sheltered inland fjords give way to a coastal archipelago with a more wave-exposed and open coast on the outside (Figure 1.2). The topography of the coast of southwestern Norway represents typical characteristics of fjord biotopes, consisting predominantly of rock, with a few areas with sand beaches. The landscape and seascape was formed mainly by the activity of the large ice sheets during the glacial periods. As a result of glacial activity, the coast is divided by a number of large fjords. Many fjords are deeper than adjacent sea areas. For example, the Norwegian Sognefjord reaches a depth of 1308 m, significantly deeper than the offshore continental shelf.

This complex topography creates barriers to the passage of water as well as organisms, meaning that a fjord system can contain several distinct habitats or even ecosystems. Fjords are situated in the cold temperate parts of the world, meaning that they are subjected to strong seasonality, with seasonal differences in light and temperature conditions between winter and summer. Seasonal differences in water temperature are strongest in the surface layers. During winter, increased mixing of surface and deeper water layers creates a uniform water column, while in the summer, increased temperature and freshwater runoff create a distinct surface layer with different water properties than the deeper layer. In periods with high freshwater runoff, such as during the snowmelt period in spring and early summer, a clear salinity gradient from the outer to the inner part of the fjord is often apparent. Surface waters, the uppermost 10–15 m, in the

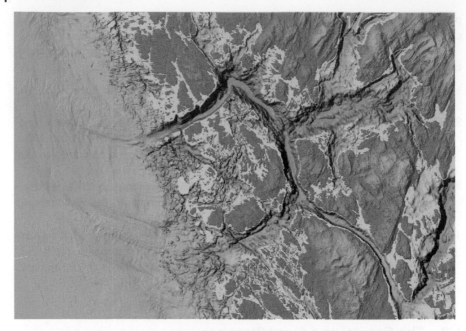

Figure 1.2 Part of the coastline of Western Norway, showing the complex bathymetry characteristic of fjords and many other coastal systems. *Source:* Gavin Macaulay, Institute of Marine Research, Norway, with permission from The Norwegian Mapping Authority.

inner parts of the fjord will have the lowest salinity. The bottom waters will often have low to zero oxygen concentrations because of the limitations sills and inlet passages set on water passage in the deeper layers; effects may be seasonal or year round. Nutrient runoff from adjacent land areas will also contribute to decreased oxygen concentrations in deep water.

The high variability in physical factors and barriers to propagation mean that fjord systems are typically home to many different communities and have an overall high biodiversity. In many cases, isolated relict populations of species can survive in fjord basins long after they have disappeared from adjacent sea areas.

1.2.1 The Littoral and Sublittoral Habitats

The **littoral zone** is used as a somewhat arbitrary term which normally refers to the intertidal and the very shallowest parts of the sea (*litus, litoris* (Latin) means "shore"). It is most commonly used for marine habitats and covers the **intertidal zone** (the area alternately covered with water or exposed to air during a tidal cycle) and the **splash zone** above the intertidal zone.

The different parts of the littoral zone can be defined by the upper and lower limits of specific zone-forming organisms (Figure 1.3). In temperate areas, the limit between the littoral and the sublittoral zone can be defined by the upper limit of kelp beds. Sometimes the shallowest part below the intertidal zone (the sublittoral zone) is also included in an expanded definition of the littoral zone. The lower extension of the sublittoral zone is arbitrary, but it is common to

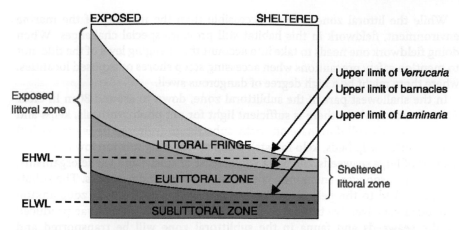

Figure 1.3 The three "universal zones" recognized by Lewis (1961), showing the zones in a gradient from extreme shelter (right) to strong wave-exposure (left). The area between the dotted lines is the intertidal area between extreme high and low water level (EHWL, ELWL). The littoral zone is composed of the splash zone (littoral fringe) and the eulittoral zone, where the limits are set by zone-forming organisms. The littoral zone is defined as the area inhabited by organisms influenced by the tidal cycle and is separated from the sublittoral zone. The width of the littoral zone is strongly extended towards the wave-exposed side due to higher waves and a much wider splash zone. In addition, the limits are shifted upwards on the wave-exposed side due to the more-or-less constant wave action. *Source:* Lewis (1961), figure modified.

separate between an upper zone inhabited by macroalgae (especially kelp) and a lower zone inhabited mostly by invertebrate animals.

The tidal range at a site is not constant, but varies throughout the moon cycle. The tidal ranges are largest at full and new moons since gravitation forces of the moon and the sun work together during these moon phases. Around full and new moons, the tides are called spring tides. When the moon is half full, the tide levels are at their smallest during the moon cycle and are called neap tides. Further, the maximum tidal range at a site is not the same during all spring or neap periods. Tidal levels along coasts are calculated and predicted by using computer models taking, among others, local topography and moon and sun cycles into account. Finally, the local tidal levels are influenced by temporary and unpredictable forces, such as air pressure, and by local weather. Strong wind toward land may, for instance, cause the sea level to rise locally. Such forces may produce variations in water levels even at sites where the local topography results in no amplitude of the tidal wave (amphidromic points).

The height of the tidal range varies between areas. Examples of places where exceptionally large tidal ranges are found are the English Channel and the Bay of Fundy (eastern Canada). Along the Norwegian coast, the range increases north-wards, from very small ranges in the Skagerrak area to spring high tide levels of more than 2 m above Chart Datum in Northern Norway. On the southwestern coast of Norway, the middle spring high tide is around 1.5 m above Chart Datum. Chart Datum is the reference level for sea maps; in most areas of Norway, it is at the same level as the Lowest Astronomical Tide level (LAT).

While the littoral zone is more accessible than the majority of the marine environment, fieldwork in this habitat still provides special challenges. When doing fieldwork one needs to take into account the changing level of the tide, not to mention safety precautions when accessing steep shores or exposed localities, where there might be a high degree of dangerous swell.

In the shallowest part of the sublittoral zone, down to around 30 m in clear Atlantic water, where there is sufficient light for net photosynthesis, kelps and other seaweeds will dominate on rocky substratum (referred to as macroalgal communities, kelp beds, kelp forests). Deeper, where light is too low to sustain growth of large seaweeds, the biota will consist of sessile and vagile organisms, for example, sponges, anemones, corals, sea urchins, and mollusks. The sublittoral is linked to the pelagic habitat and receives nutrients and zooplankton produced via advective transport. Spores, gametes, eggs, and larvae produced by the seaweeds and fauna in the sublittoral zone will be transported and dispersed through the pelagic habitat.

In temperate areas, the sublittoral kelp beds have an extraordinarily high primary production, biodiversity, and density of invertebrates (Mann, 2000). Studies of invertebrates associated with different aquatic plants and seaweeds have shown numbers of up to several hundred thousand individuals m^{-2} per seaweed vegetation (Christie *et al.*, 2009). Since many of the small invertebrates are important as food for larger organisms, the kelp beds form an important habitat for a number of fish species, like labrids and small gadoids, and provide both food and shelter for a number of fishes and large invertebrates. These areas are often important nursery areas for a number of species.

1.2.2   The Continental Shelf and Slope

Continental shelves form a narrow fringe around the continents that vary in width, making up around 7% of the ocean's surface and <0.2% of its volume. Continental shelves vary in both width and depth down to 200 m. Environmental and biological characteristics of shelf areas depend on land impact, depth, steepness, seabed texture and topography of the shelf, and the ocean and currents along the slope side. Vertical stability is lower on the shelf compared to the ocean floor. This enhances nutrient availability and thus primary production by algal cells that, in turn, also increases food availability and production at higher trophic levels (Postma and Zijlstra, 1988). The continental shelf gradually gives way to the continental slope, which serves as the transition to the abyssal plains of the deep sea. On the continental slope, submarine canyons created by flows of sediment from the shelf often serve as unique and diverse habitats for benthic organisms.

Upwelling ecosystems are shelf areas that are particularly productive due to continuous wind-induced upwelling of nutrient-rich deep water, often lying over seabeds rich in fine sediments. The nutrient-rich waters then stimulate growth and production of plankton, which has a cascading effect on higher trophic levels. The best known upwelling ecosystems are the Benguela Upwelling System (on the west coast of Namibia and South Africa), California Current

System, Canary Current System, Humboldt Current System, and the Somali Upwelling System (See Mann and Lazier, 2006 for details).

The **North Sea** is an example of a relatively large continental shelf, also considered a semi-enclosed sea (Figure 1.1). The North Sea contains many shallow bank areas in the south and more channels, deep holes, and slope areas in the north. In the southeast, the North Sea is bordered by the European continent. Here it is shallow and has large areas of mud, most of which are in the estuarine area called the Wadden Sea. In the northeast, the shelf area drops into a deep trench, the Norwegian Trench, and is bordered by the Scandinavian peninsula. To the west and north, the North Sea is bordered by the British Isles, with the English Channel connecting it to the Atlantic, and the deeper Norwegian Sea (see Zijlstra, 1988 for further details).

1.2.3 The Deep Ocean

The seafloor of the world's major oceans is created by the action of divergent continental plates. Because of the relatively high weight of the basalt-rich rocks of the seafloor crust when compared to granite-rich rocks of continental landmasses, the depth of the major ocean basins is usually around 4000–5000 m. The deep sea can be defined in a few different ways, but because changes in temperature are much less pronounced at depths below 1500–2000 m, depths in this range are often used to distinguish the deep sea from shallower areas. Thus the deep sea comprises part of the bathyal or bathypelagic zone (1000–4000 m), the abyssal or abyssopelagic zone (4000–6000 m), and the hadal zone (>6000 m). Hadal depths are usually common for special geological features such as back-arc basins, created by the break-up of tectonic plates due to nearby subduction zones.

The deep-sea seabed is uniform in physical characteristics over large areas, typically has low amounts of available energy and nutrients, but is home to a surprisingly high diversity of benthic organisms. Special features, such as mid-ocean ridges, volcanic islands, and seamounts, are found in the deep-sea environment and represent distinct ecosystems. As a special case, hydrothermal vent systems along mid-ocean ridges support rich communities based on chemoautotrophic bacteria living on enriched vent fluid.

The deep sea is characterized by several gradients in abiotic factors, such as pressure, light (lacking in the aphotic zones), temperature, and dissolved oxygen. Most importantly, the availability of nutrients rapidly diminishes with distance from photosynthetic primary production at the ocean surface. Early views considered the deep sea as mostly devoid of life or characterized by more primordial organisms.

Surveys of life in the deep sea have shown that while biomass is lower, there is an extraordinary diversity of organisms, with a maximum at the mid to lower bathyal zone (Rex and Etter, 2010). While there are several hypotheses regarding this surprising diversity, the comparative lack of experimental data means that these are somewhat tentative. The environmental stability hypothesis is based on the relative unchanging conditions of the deep sea, while the intermediate disturbance hypothesis explains diversity in terms of the "right" amount of disturbance present. The sheer size of the deep sea may facilitate speciation. Food particle diversity is greater at depths >1500 m, which may lead to further specialization and, for benthic organisms, the small size of benthic detritivores means that there is further room for exploitation of microvariations in benthic topography.

1.3 Physical Characteristics of the Pelagic System

Dag L. Aksnes, University of Bergen and Lars Asplin, Institute of Marine Research, Norway

Here, we make a brief account of some of the physical variables that structure the pelagic ecosystem. Our emphasis is on the difference between oceanic and coastal locations that are affected by freshwater supply. For example, the Norwegian coast is surrounded by Norwegian Coastal Water (NCW). NCW is characterized by lower salinity (Figure 1.4, yellow and green shading) than the oceanic water, denoted North Atlantic Water (NAW, salinity > 35 ppt). Strong gradients are observed in the environmental variables when moving offshore from the coast toward the ocean (Sætre, 2007). Here, this is illustrated by an idealized transect from inshore to offshore Norway (Figure 1.5, Figure 1.6 and Figure 1.7). The lower salinity (<34.5 ppt) of NCW is caused by large freshwater drainage from the Baltic Sea, the southern North Sea, and local drainage along the Norwegian coast. Due to the rotation of the earth (e.g., the Coriolis Effect and Ekman drift), NCW is like a giant "river", pressing against the coast on its northward movement toward the Barents Sea. This current is named the Norwegian Coastal Current (NCC). West of the NCC we have the North Atlantic Current (NAC), which is a northern branch of the Gulf Stream, carrying saline (>35 ppt) NAW into the Norwegian Sea.

We now consider an imaginary cruise line from the innermost part of a coastal location (e.g., a fjord), through the coastal water and out into the open ocean (Figure 1.5a). At the head of the fjord, a river drains into the sea; this gives rise to a relatively local, thin, brackish layer. This is because freshwater is less dense (i.e. weighs less) than saline water. Moving outwards toward the ocean, the brackish water is gradually mixed into the underlying NCW and the brackish layer will soon disappear. Also further out, the NCW is mixed with the underlying NAW and transported northwards along with the NCC. Thus, at some point along the transect, the NCW disappears and is replaced by NAW.

A water column, which is characterized by increasing density with depth, is said to be density stratified. This is an important concept in oceanography. A stratified water column requires more energy from wind, waves, and tides, for example, to be vertically mixed. Although the density of seawater is controlled by both temperature and salinity, salinity has the strongest effect on density stratification in areas with large freshwater influence, such as where NCW is present. In the open ocean, where vertical salinity gradients are small, density stratification is mainly controlled by the vertical gradient in temperature.

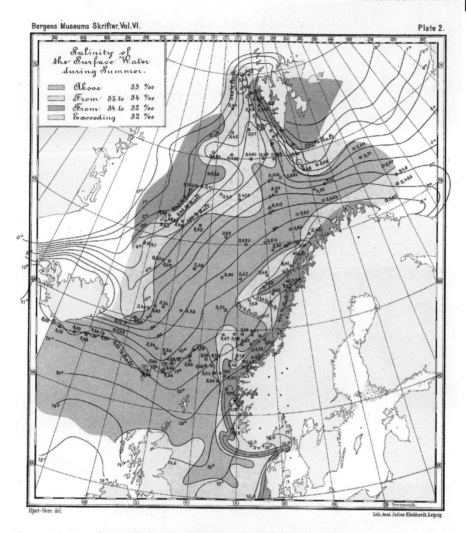

Figure 1.4 An early map illustrating the Norwegian Coastal Water (yellow and green) between the Norwegian coastline and the oceanic North Atlantic Water (blue) (Hjort and Gran, 1899). *Source:* University of Bergen Library, Norway.

We now consider an idealized "steady state" summer situation where the weather has been calm for a period of time (Figure 1.5b, 1.5c, and Figure 1.6). The strong density stratification at the coastal location results in a higher surface temperature than at the oceanic location (Figure 1.5b and 1.6b). At both locations, sun radiation is quickly absorbed and converted to heat in the upper few meters. Due to the weaker density stratification (small vertical gradient in salinity) at the oceanic location (Figure 1.6c), vertical mixing will be higher than at the coastal location. This mixing will result in a larger net transport of heat downwards and consequently leads to a lower surface temperature than at the coastal location.

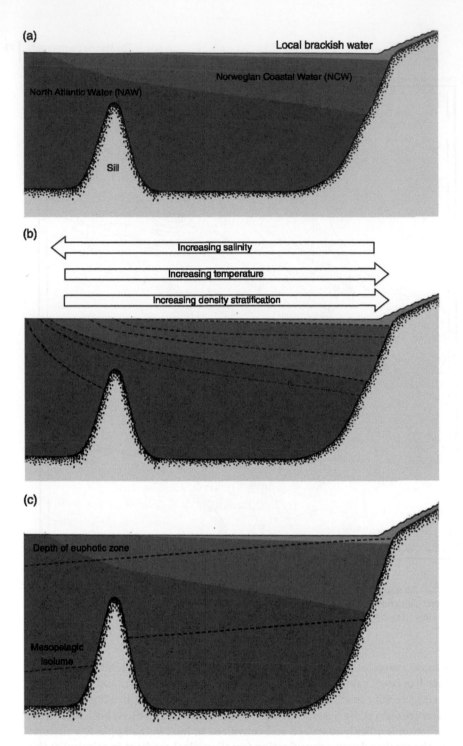

Figure 1.5 A transect from the head of a fjord (right) to an oceanic location (left) showing (a) the three main water masses: local brackish water, Norwegian Coastal Water (NCW), and North Atlantic Water (NAW). A summer situation is indicated in (b). The arrows indicate the gradients in salinity, temperature, and density stratification; broken lines indicate isoclines for the three variables. Fresher water contains higher concentrations of CDOM with terrestrial origin causing isolumes, like the euphotic depth, to shoal toward locations more affected by freshwater. Depth of euphotic zone and Mesopelagic isolume is indicated in (c).
Source: Artwork by R. Jakobsen.

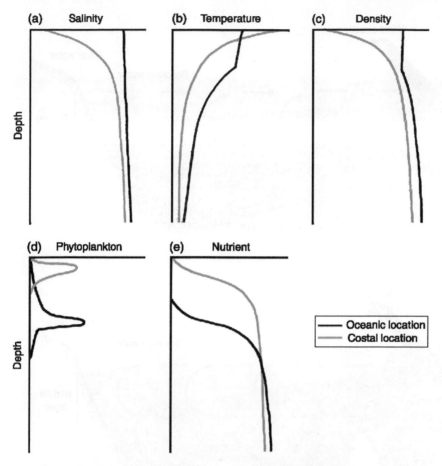

Figure 1.6 Difference in vertical distributions of (a) salinity, (b) temperature, (c) density, (d) phytoplankton, and (e) nutrients at a coastal (grey) and an oceanic location (black) in an idealized steady state summer situation. *Source:* Artwork by R. Jakobsen.

Freshwater affects light penetration. This is because freshwater drainage from land contains a high concentration of dissolved organic matter (DOM). These substances are referred to as chromophoric or colored dissolved organic matter (CDOM). A large fraction of the CDOM, which is mainly of terrestrial origin (e.g. humic substances), is resistant to microbial degradation and has long residence times in the ocean. CDOM is a strong light absorber and therefore increases the attenuation of sunlight on its way through the water column. The CDOM light attenuation tends to be proportional to the fraction of the freshwater and consequently relates inversely to salinity (Stedmon and Markager, 2003). Thus NCW attenuates light more than NAW (Aksnes, 2015). This means that the depth of the euphotic zone, which corresponds to around $1-10\,\mu$mol quanta m^{-2} s^{-1}, is shallower in coastal than in oceanic water (Figure 1.5c). Such shoaling of isolumes (Figure 1.5c) not only has consequences for the vertical distribution of photosynthetic organisms like phytoplankton (Figure 1.6d) and benthic algae, but also of inorganic nutrients (Omand and Mahadevan, 2015, Figure 1.6e) as

(a)

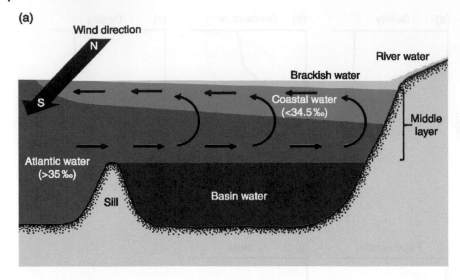

(b)

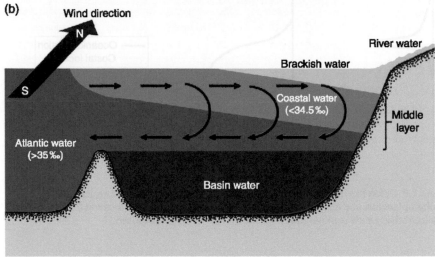

Figure 1.7 Effects of coastal winds on circulation patterns of coastal areas and fjords. Northerly and southerly winds cause (a) upwelling and (b) downwelling with opposite circulation patterns in the intermediate layer (the water layer between the brackish water and the sill depth) of a fjord. *Source:* Artwork by R. Jakobsen.

well as of organisms that make use of light in vision and orientation. For example, a mesopelagic organism searching for a light comfort zone of the order 10^{-5} µmol quanta m^{-2} s^{-1} will generally find this zone several hundred meters shallower in a murky coastal than a clear oceanic location (Røstad, Kaartvedt, and Aksnes, 2016). Similarly, we would expect that the amplitude of diel vertical migrations would be less pronounced in murky coastal than in clear oceanic water.

It should be noted that a number of processes not considered here affect the patterns discussed above, for example, coastal upwelling, downwelling, storm events, and mesoscale eddies will all affect these patterns (Sætre, 2007). We make one remark concerning episodic downwelling and upwelling events that occur along the coast as a function of wind patterns. With persistent wind from the north along the coast, the NCW tends to drift toward the ocean due to Ekman drift (outward directed arrows in Figure 1.7a). This loss of coastal water is compensated by deeper water that moves in the opposite direction (inward and upward arrows in Figure 1.7a) a process termed upwelling. With wind from the south, the situation is reversed (Figure 1.7b). The NCW now presses against the coast and causes a downwelling situation. Thus the wind situation at the coast has a large effect on the circulation patterns and the advection of, for example, zooplankton in coastal areas and fjords (Aksnes *et al.*, 1989). The water transport into (and out of) a fjord in the intermediate layer typically amounts to thousands of m^3 per second, while the estuarine circulation (not illustrated in Figure 1.7) that is associated with local river discharges is an order of magnitude lower.

While the deep ocean is generally well oxygenated due to the global thermohaline circulation, this is often not the case in coastal locations, for example, the deep basins of the Baltic Sea and the Black Sea are permanently anoxic. This is because shallow sills hamper the exchange of the basin water with outer, oxygen-rich water. When the basin water has a long residence time it becomes stagnant; oxygen is consumed more rapidly than it is supplied, resulting in anoxia. Several fjords are similarly characterized by low concentrations of dissolved oxygen and even anoxia. Generally, this occurs when the sill is shallow so that the basin water of the fjord has no direct contact with the outer oceanic water. At times, for example, with long periods of northerly wind and upwelling of dense deep water along the coast, dense surface water might intrude into the fjord above the sill depth and then sink into the basin due to its greater weight. Such renewals of the basin water will lead to a temporary increase in the oxygen concentration.

1.3.1 The Light Environment

Dag L. Aksnes, University of Bergen

The variation in light intensity during a 24 h day–night cycle or over a 1000 m water column typically spans 10 orders of magnitude. Such huge temporal and spatial variations are seldom seen for other environmental variables. Thus it is no surprise that organisms respond to changes in light such as in diel vertical migration. Measuring the light environment is less trivial than measuring salinity, temperature, and oxygen for example. Some knowledge of optical quantities and light transmission are required to know what different instruments measure.

1.3.1.1 Inherent Optical Properties: Scattering and Absorption Coefficients

The common definition of light is the electromagnetic radiation that can be sensed by the human eye. Visible light spans approximately from 390 to 700 nm. It is sometimes convenient to sum the energy (photons) in this band, but since the photons of different wavelengths are **absorbed** and **scattered** differently, this is not always appropriate. As the photons travel through water, some are **absorbed** (a), that is, converted to heat, and some are **scattered** (b), which means that the direction of the photon is changed. These two **inherent** optical properties of the water have units per meter (m^{-1}) and determine how fast light is attenuated in water. An absorption coefficient of, for example, $0.3\,m^{-1}$ for a particular wavelength means that 74% $(e^{-0.3} \times 100\%)$ of the photons of this wavelength "survive" per meter travelled distance and 26% are absorbed, that is, are converted to heat. Similarly, a scattering coefficient of $0.3\,m^{-1}$ means that 26% of the photons change direction and that 74% of the photons of the wavelength in question pass through one meter in a straight line.

1.3.1.2 Visibility, Sighting Distance, and the Beam Attenuation Coefficient

Light is attenuated more in water than in the atmosphere because water absorbs more photons than air. In addition to absorption, light is also scattered more in water than in air (unless it is very foggy). A light beam (photons moving in the same direction) is destroyed as it moves through water because the photons are scattered in different directions. Visibility is based on image transmission, that is, a light beam that preserves the image through space/water. Scattering (in addition to absorption) therefore destroys the image of an object along the path of sight. That is why the underwater sighting distance is short in water (often a few cm in particle rich water, but more than 50 m in very clear water) compared to the atmosphere.

It is important to recognize that there are two different light attenuation coefficients. Most ecological textbooks refer to the "attenuation of light" between the surface and a specific depth (such as the compensation depth or the euphotic depth). We return to this in Section 1.3.1.3, but now consider the attenuation relevant for transfer of images, that is, light beams. The **beam attenuation** coefficient, given the symbol c with unit m^{-1}, tells us how rapidly a light beam, and consequently an image, is destroyed. Since the image of a copepod, for example, is carried through the water by photons travelling in straight lines from the copepod to the eye of the fish, the beam attenuation coefficient determines the maximal **sighting distance** in water. The beam attenuation is simply the sum of absorption (a) and scattering (b), that is, $c = a + b$. A high c means poor **visibility** and a short sighting distance. The maximal sighting distance depends also on the contrast of the object and the sensitivity of the visual system of the viewer. Remember also that a, b, and c are all wavelength specific and so are sighting distances, contrasts, and visual sensitivity. Particularly in shallow water with a large span in the wavelengths of incoming sunlight, the wavelength composition is important for vision. As the downwelling sunlight becomes more monochromatic at larger depths (i.e. only photons of a narrow wavelength band around 480 nm have survived), the situation becomes somewhat simpler (see Figure 1.8).

1.3.1.3 Light Penetration and the Attenuation Coefficient of Diffuse Light

Clear water (i.e. a low c) is not sufficient to see an object. Obviously, without light, nothing can be seen. If sunlight is utilized in vision, the sunlight intensity at depth must be sufficient. The **attenuation coefficient of diffuse light**, or more precisely the **attenuation coefficient of downwelling irradiance** (K, also with the unit m^{-1}) determines how large a fraction of the surface light penetrates down to a particular depth. Like the beam attenuation coefficient, K is also a function of a and b, but K is generally much lower than c. This is because sunlight photons at a particular depth might not have travelled in a straight line, but

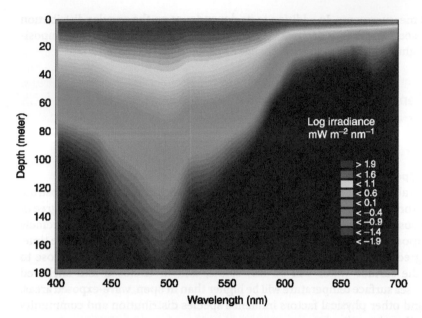

Figure 1.8 Measurements of downwelling irradiance with a spectroradiometer in the Norwegian Sea. Note that most of the light that remains at large depth narrows to around 480 nm. *Source:* D.L. Aksnes.

might have changed direction a number of times before they reached the depth in question. A flat sensor facing upward is used to measure downwelling irradiance. Thus all photons from above, regardless of their hitting angle, are counted. If the same sensor faces downward, the upwelling irradiance is measured (photons that have been scattered upwards). We now understand why the attenuation coefficient of downwelling irradiance, K, is useful to characterize the amount of sunlight that is available for vision or photosynthesis at a particular depth, while the beam attenuation, c, is useful to characterize the visibility between the object and the viewer. Again, it should be noted that K, like c, varies with the wavelength. Furthermore, K is said to be an **apparent optical property** of water because it depends not only on the properties of the water (i.e. inherent to the water), but also on the angle of the incoming sunlight (the angular distribution of light). For example, if the sun is at its zenith, the photons have on average a shorter distance to travel to a particular depth than if the sun is just above the horizon. This means that K for a particular water column is lower at midday than during dawn and dusk.

1.3.1.4 Photosynthetically Active Radiation (PAR)

Photosynthetically Active Radiation (PAR) is the summed energy in the wavelength band 400 to 700 nm. This quantity is often referred to in textbooks and used in ecological studies. PAR measurements are popular because they provide just one number for the light intensity (i.e. wavelength distribution is ignored) and robust sensors are commercially available. Note, however, that the rationale for the PAR definition is photosynthesis and not vision. PAR is a good predictor for the photosynthetic rate, but there are some caveats in using K's which are based on PAR measurements. In addition to its dependence on the angular distribution of light (see 1.3.1.3), PAR derived K's are also affected by the wavelength composition of light and this composition changes with depth (see Figure 1.8).

1.4 Temperate Marine Communities – Environment and Organisms

1.4.1 Littoral Organisms

1.4.1.1 Species, Zonation, and Communities

The littoral zone does not form a uniform biotope within an area. The substrate in the littoral zone may vary from solid rock to muddy substratum. The degree of wave exposure will vary from perfectly calm and sheltered estuaries to extremely wave-exposed sites on the open coasts, subjected to heavy swell most of the time. The degree of inclination may also vary substantially along the coast. Close to rivers, the surface water will be less saline, and in enclosed and sheltered bays, the sea surface temperature will be higher than in open, wave-exposed areas. These and other physical factors influence species distribution and community composition along the shore.

The species composition and type of community found at a particular littoral locality is influenced by both abiotic and biotic factors. The most important abiotic factors are **substratum type** (rocky vs. soft) and **degree of exposure to waves** (sheltered vs. exposed). Biotic factors like **competition** for substrate and sunlight, **trophic interactions** (predation), and **anthropogenic disturbance** (recreational use or pollution) are further important in shaping the littoral community.

Macroalgae and sessile animals require firm substratum and are consequently excluded from littoral areas composed of sand or mud. In such habitats, animals that may live buried or dig through the substratum will dominate. Strong waves create a harsh physical environment in the intertidal and upper sublittoral zone of strongly wave-exposed areas, and has a big impact on the biota in such places. The intertidal zone tends to be dominated by either small, sessile animals and either small and turf forming or long and flexible macroalgae (Figure 1.9). In the North Atlantic, sessile animals like the barnacle *Semibalanus balanoides* and the blue mussel *Mytilus edulis* commonly dominate. In addition, long and flexible kelps, like *Alaria esculenta* and *Laminaria digitata*, dominate in the lower intertidal zone.

Sheltered rocky shores of the temperate North Atlantic are typically dominated by large brown macroalgae (Figure 1.10). Along most of the Norwegian coast, the following intertidal zonation pattern is characteristic for sheltered sites:

1) In the uppermost part of the littoral zone, we typically find encrusting lichens and, a bit lower, tufts of the brown alga *Pelvetia canaliculata*.
2) In the mid–upper intertidal zone, a common dominating brown alga is *Fucus spiralis*, with *Fucus vesiculosus* occurring lower. *F. spiralis* is typically found at somewhat more sheltered localities, while *F. vesiculosus* may dominate in more wave-exposed areas. A narrow zone of the barnacle *Semibalanus balanoides* can be found above or between the *Fucus* zones.

Figure 1.9 The intertidal community at a locality with high degree of wave exposure. The shore is dominated by barnacles and small, turf-forming algae. *Source:* K. Sjøtun.

Figure 1.10 The intertidal community at a sheltered locality with low degree of wave exposure. The shore is dominated by large brown algae. *Source:* K. Sjøtun.

3) In the mid–lower intertidal zone, the most common dominating species is the brown alga *Ascophyllum nodosum*. The amount of *Ascophyllum* is related to degree of wave exposure and inclination of the locality.

4) Below *A. nodosum* there will normally be a zone of *Fucus serratus*.

5) In the sublittoral zone, temperate kelps dominate, with *Saccharina latissima* being most common in sheltered sites and *Laminaria* in more wave-exposed sites.

The horizontal bands formed by some dominating organisms are especially conspicuous in sheltered sites. These bands or zones are mainly shaped by a combination of differential tolerance to physical stress by the organisms and competition between them. Within the littoral zone, the ebb and flow of the tide creates a gradient in the amount of time the organisms are exposed to air versus the amount of time they are submerged under water. Organisms growing in the lowermost part of the intertidal zone will spend only a relatively short time exposed to air, while the organisms in the upper part must tolerate being out of seawater for many hours per tidal cycle. In this way, the tidal cycle creates a **gradient of physical stress** to the marine organisms inhabiting this zone, which is a major feature of this habitat. The organisms we find in the intertidal zone are specialized so that they tolerate this environment and most of them only live in the intertidal biotope. Relatively few organisms have adapted to the stressful conditions of the intertidal; normally species richness decreases toward the upper part of the intertidal level, where the period of air exposure is highest.

Typically, many dominating organisms in the littoral zone inhabit different niches and are specialized toward different levels of air exposure. The **physiological tolerance levels** toward, for example, desiccation, temperature, and salinity determine at which vertical level a species can survive. This, in combination with **competition between habitat forming species**, produces characteristic patterns of **vertical zonation**, where different algae and animals typically form dominating and distinct bands at fixed vertical levels along the shore.

Soft-bottom littoral communities differ significantly from their hard substratum counterparts. Beaches and other soft-bottom littoral zones usually have a very shallow inclination. Lack of suitable substrate means that large macroalgae or encrusting organisms are normally not found. Instead, we find a variety of burrowing animals, such as different species of polychaetes, crabs, and bivalves. While briefly mentioned here, intertidal soft-bottom communities will not be treated further.

1.4.2 Sublittoral Organisms

The sublittoral zone is highly productive and houses a rich flora and fauna. In the North Atlantic, large extensions of kelp, mainly of the genera *Laminaria* and *Saccharina*, serve as sources for primary production in the nearshore areas and as substrate for sessile benthic organisms and for epiphytes of algae. In addition, kelps, together with other large habitat-forming seaweeds in the coastal zone, form a three-dimensional shelter for numerous small invertebrates, with

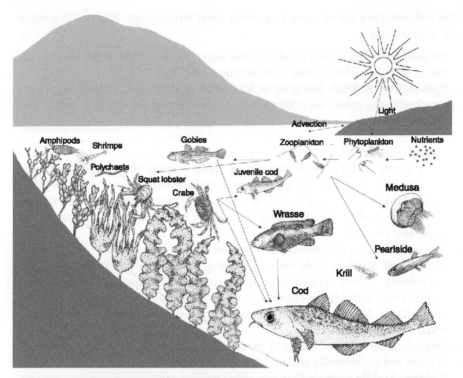

Figure 1.11 A typical sublittoral habitat illustrating trophic relations and links to the physical environment. *Source:* Artwork by R. Jakobsen.

gastropods and amphipods being most abundant. Many nest-building fish species reproduce here and utilize seaweeds as substrate for their nests. Examples are members of the families Gobiidae (gobies) and Labridae (wrasses), where the males build nests for the females to deposit their eggs. The males care for the eggs until hatching. These fish typically stay in the sublittoral during their entire life cycle. Juveniles of other species settle in the sublittoral before moving deeper with age. Examples from temperate areas of the North Atlantic include coastal populations of species from the family Gadidae (codfish). Gadoids have pelagic reproduction and pelagic eggs and larvae. The sublittoral provides shelter and feeding opportunities for the post-settlement stages. As juveniles grow larger, their feeding and shelter needs change and they migrate to deeper waters. An example of a temperate sublittoral habitat in a North Atlantic fjord and links to the pelagic and outer coast is shown in Figure 1.11.

1.4.3 Demersal and Benthic Organisms

1.4.3.1 Bottom-associated Organisms
Organisms associated with the seabed can be defined in a variety of ways depending on which characteristics are deemed most important in a particular setting.

The following is a list of some commonly used terminology describing bottom associated fauna:

- **Demersal**: Animals associated with the water column close to the seabed (the demersal zone). Mostly used in connection with fish.
- **Epibenthos**: Animals living on or immediately above the seafloor. Some are attached to the substrate, others are mobile. Examples are sponges, corals, and sea stars.
- **Endobenthos = Infauna**: The animal life within sediment.
- **Hyperbenthos**: Small-sized bottom-dependent animals that have good swimming ability and perform seasonal or daily vertical migrations above the seabed, with varying amplitude, intensity, and regularity.
- **Hypoplankton**: Forms of marine life whose swimming ability lies somewhere between that of the plankton and the nekton; includes some mysids, amphipods, and cumacids.
- **Macrofauna**: Animals visible to the naked eye.
- **Meiofauna**: Small benthic animals ranging in size between macrofauna and microfauna. Often defined as animals not retained by a 1 mm mesh size; includes interstitial fauna (animals living in between sediment particles).
- **Microfauna**: Microscopic animals such as protozoa (unicellular animals) and small nematodes.
- **Nektobenthos**: Those forms of marine life that swim just above the ocean bottom and occasionally rest on it.
- **Nekton**: Mobile animals that can swim against currents, Examples are fish, cephalopods, large crustaceans, and even whales.

1.4.3.2 Continental Shelf and Slope Benthos

The continental shelf and slope is a vast habitat containing a range of sub-habitats associated with differences in substrate composition and other abiotic factors. A large part of the continental shelf and slope seafloor is composed of soft-bottom sediments. While variable in qualities, it is nevertheless possible to extract a few generalities that characterize most soft-bottom sediments, namely that they are dominated by invertebrates more than 0.5 mm in length and that these belong to a few characteristic phyla:

1) Polychaeta: mostly suspension-feeding, deposit-feeding, or predatory;
2) Crustaceans: mostly scavengers, suspension-feeders, grazers, and predators;
3) Echinoderms: suspension-feeders, grazers, and predators;
4) Cnidarians: suspension-feeders and/or predators;
5) Mollusks: all feeding modes.

Latitude also appears to have a structuring effect on the soft sediment communities. In polar regions, which constitute about 25% of the world's open ocean areas and 25% of the world's continental shelf areas, communities are predominantly:

1) Suspension feeders: dominated by sponges, clams, polychaetes, and soft corals;
2) Deposit feeders/grazers: mainly sea stars, sea urchins, polychaetes, and isopods;
3) Predators: fish, amphipods, pycnogonids, sea stars, amphipods, and polychaetes.

Meanwhile, the temperate zone, which includes 35% of world's open ocean areas and about 45% of world's continental shelf areas, the soft sediment communities consist of:

1) Suspension feeders: dominated by clams and polychaetes;
2) Deposit feeders: mainly clams and sea cucumbers;
3) Predators: crabs, amphipods, gastropods, and polychaetes.

1.4.3.3 Benthic Fish of the Continental Shelf and Slope

Continental shelf areas are very productive areas where most fisheries occur. This is also the reason why both fundamental and applied research have a long history in continental shelf areas; basic research questions, such as "Why do fish stocks fluctuate?", started in the nineteenth century. The North Sea ecosystem and other continental shelf ecosystems of the world are described in Postma and Zijlstra (1988), while details of the Norwegian Sea Ecosystem can be found in Skjoldal (2004). An overview of all species can be found in Wheeler (1969), Moen and Svensen (2004), and Heessen, Daan, and Ellis (2015), but see also the web page at the University of Bergen and Institute of Marine Research, Norway.

In **shallow, nearshore coastal areas**, small species of the families Gobiidae (gobies) and Labridae (wrasses) are common and very abundant, particularly during the summer season when they reproduce. On the west coast of Norway, the most numerous small fish species is the two-spot goby (*Gobiusculus flavescens*; Fosså, 1991) and five labrid species; rock cook (*Centrolabrus exoletus*), goldsinny (*Ctenolabrus rupestris*), ballan wrasse (*Labrus bergylta*), cuckoo wrasse (*Labrus bimaculatus*), and corkwing (*Symphodus melops*) (Salvanes and Nordeide, 1993). In addition, juvenile cod (*Gadus morhua*), pollack (*Pollachius pollachius*), and saithe (*Pollachius virens*) have their nursery area here, while the small gadoid species, poor-cod (*Trisopterus minutus*), are also abundant (Salvanes and Nordeide, 1993).

The fish fauna of the North Sea and surrounding coastal areas is rich and consists of at least 160–170 species (see Heessen, Daan, and Ellis, 2015). Those that are fished upon are best known and belong to the families Gadidae, Ammodytidae, and Pleuronectidae, which include cod (*Gadus morhua*), Norway pout (*Trisopterus esmarkii*), whiting (*Merlangus merlangus*), saithe (*Pollachius virens*), sand-eel (*Ammodytes* spp.), and plaice (*Pleuronectes platessa*). Rays and skates (Rajiformes) are also common on the shallow banks of the North Sea.

The benthic shelf-edge and slope fishes. On the continental shelf and slope of the eastern Norwegian Sea, a total of sixty fish species have been described. Biomass and diversity decrease with depth (Bjelland and Holst, 2004). The fish communities are divided into four categories, characterized by environmental variables and depth (Bergstad, Bjelland, and Gordon, 1999; Bjelland and Holst, 2004; Figure 1.12):

1) **The deep-water species** consist mainly of eelpouts and snailfish (*Lycodes frigidus, Paraliparis bathybius*, and *Rhodichthys regina*).
2) Common "upper-slope cold species" are rays (*Raja hyperborea* and *R. clavata*), the two eelpout species (*Lycodes pallidus* and *L. flagellicaudata*), and the gadid Arctic rockling (*Onogadus argentatus*).

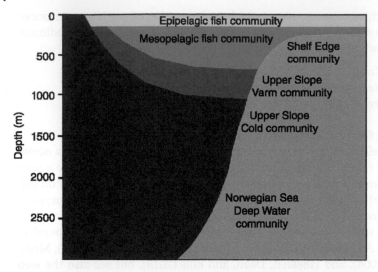

Figure 1.12 Classification of fish communities by depth using the Norwegian Sea shelf slope summer situation as an example. *Source:* Redrawn from Bjelland and Holst (2004). Artwork by R. Jakobsen.

3) The most common "upper-slope warm species" include four eelpout species (*Lycodes esmarkii*, *L. eudopleurostictus*, *L. seminudus*, and *Lycenchelys muraena*), Greenland halibut (*Reinhardtius hippoglossoides*), roughhead grenadier (*Macrourus berglax*), and the bullhead (US: sculpin; *Cottunculus microps*).
4) The most important "shelf-edge species" include redfishes (*Sebastes* spp.), ling (*Molva molva*), tusk (*Brosme brosme*), and monkfish (*Lophius piscatores*).

1.4.3.4 Deep Bottom Fish of Fjords and the Norwegian Deeps

In Masfjord at depths >400 m, the dominating bottom fish species are roundnose grenadier (*Coryphaenoides rupestris*); three chondrichthyan species: black-mouthed dogfish (*Galeus melastomus*), velvet belly (*Etmopterus spinax*), and rabbit fish (*Chimaera monstrosa*); blue ling (*Molva dipterygia*): tusk (*Brosme brosme*); sølvtorsk (*Gadus argenteus*); witch (*Glyptocephalus cynoglossus*); European hake (*Merluccius merluccius*); and the hagfish (*Myxine glutinosa*). Saithe was also observed on the seabed from an ROV. Most of these deep bottom associated fjord species are also found in the Norwegian deeps and along the continental slope (see Bergstad, 1990a, b for details). However, the deep-water benthic fish community is more diverse outside of Masfjord (Bergstad, 1990a).

1.4.4 Pelagic Organisms

1.4.4.1 Plankton and Micronekton
Tone Falkenhaug, Institute of Marine Research, Norway

Plankton is a diverse group of organisms that spend either part or all their life drifting in the water column. Although many of these organisms are capable of

Table 1.1 Plankton size classes.

Group	Size	Examples
Femtoplankton	<0.2 μm	Viruses
Picoplankton	0.2–2 μm	Bacteria
Nanoplankton	2–20 μm	Dinoflagellates
Microplankton	20–200 μm	Diatoms, eggs and larvae of crustaceans
Mesoplankton	0.2–20 mm	Copepods, fish larvae, chaetognaths
Macroplankton	2–20 cm	Euphausiids, amphipods, jellyfish
Megaplankton	>20 cm	Large jellyfish, salps

locomotion, they generally have low swimming ability and are unable to move against currents. This separates plankton from nekton, which includes organisms that can control their movement in the water (such as fish). Plankton are generally smaller than nekton, however, some planktonic organisms can be quite large (jellyfish up to a meter or more) (Table 1.1). **Micronekton** is the name for an intermediate group between plankton and nekton; usually they are smaller pelagic organisms and caught in moderately sized trawls with mesh sizes of 4–5 mm. Micronekton, consisting mainly of decapod crustaceans, smaller cephalopods, and small fishes, are described below under "mesopelagic organisms".

Many zooplankton and micronekton undertake **diel vertical migrations** (DVM) over the course of every 24-hour period, moving into the upper waters at night and descending into the darker depths during the day. This behavior is considered an adaptation for feeding in food-rich surface waters at night and avoiding visual predators during the day. The DVM of zooplankton may span several hundred meters and is considered the largest (in terms of biomass) and most regular migratory movement on the planet.

It is possible to classify members of the plankton in multiple ways. One way is according to how they obtain energy: **phytoplankton** are plant-like autotrophs that perform photosynthesis, while **zooplankton** are animals that consume other organisms.

Size is another way to categorize plankton, which includes organisms from micrometers (viruses) to several meters (large jellyfish). Usually the following divisions are used, which reflects the mesh size needed to filter them out of the water:

Plankton may also be categorized according to how much of the life cycle is spent in the plankton. **Holoplankton** are planktonic their entire lives (e.g. phytoplankton, copepods, ctenophores, chaetognaths). **Meroplankton** spend only a part of their life cycle as plankton. This group includes organisms with planktonic larvae that eventually change into a bottom-living (worms, mollusks, crustaceans, corals, echinoderms) or free-swimming life stage (fishes).

Phytoplankton, also called micro-algae, are single-celled, photoautotrophic microorganisms. Phytoplankton account for half of the photosynthesis on earth and play an important role in the removal of CO_2 from the atmosphere.

Phytoplankton are the basis for the vast majority of oceanic food webs. Since phytoplankton are dependent on sunlight for photosynthesis, they are restricted to the photic zone in the upper 50–100 m of the water column. Growth also depends upon mineral nutrients, which is supplied from deeper layers or land. Given enough sunlight, CO_2, and nutrients, populations of phytoplankton can reproduce explosively, doubling their numbers in just one day. There are about 4000 described species of marine phytoplankton. In terms of numbers, there are four important groups of phytoplankton:

- **Diatoms** have an outer shell comprised of silica and are the ecologically most important group in arctic and boreal ecosystems, usually dominating the phytoplankton spring bloom.
- **Dinoflagellates** are more common in the summer and autumn period. Some species within this group are known to produce toxins that can be harmful for humans through accumulation in shellfish.
- **Coccolithophores** have outer shells comprised of calcium carbonate and thrive in warm, nutrient poor waters. Blooms of the coccolithophore species *Emiliania huxleyi* usually occur during summer in coastal waters and fjords, which gives the ocean a chalky color, visible from space.
- **Cyanobacteria** (sometimes erroneously called blue-green algae) are a group of bacteria that are able to perform photosynthesis. Blooms of toxin producing cyanobacteria have become an increasing problem in polluted estuarine and brackish waters, such as the Baltic Sea.

Zooplankton are consumers that eat other plankton and thus provide an important link between primary producers (phytoplankton) and higher trophic levels. Some are herbivorous, filtering phytoplankton out of the surrounding water, whilst others are carnivorous predators on smaller zooplankton. Zooplankton are found in all oceans, from the surface to the deepest trenches. Every major phylum of the animal kingdom is represented in the zooplankton, either as adults or as larvae. The following taxonomic groups are commonly found in plankton samples.

The **Copepods** are small crustaceans of great ecological importance. The most abundant copepod species in the North Atlantic, *Calanus finmarchicus*, feeds on phytoplankton and, in turn, is an important food source for fish larvae and pelagic fish (herring and mackerel). During spring, this species builds up fat reserves (omega-3 fatty acids), which it draws upon during the winter when it hibernates 2000 meters below the sea surface. **Euphausids** are relatively large shrimp-like crustaceans that are more commonly known as krill. The Nordic krill (*Meganyctiphanes norvegica*) reaches a length of 45 mm and is widespread in the North Atlantic. Krill are often found in large swarms particularly in the polar seas, providing food for whales, fish, and birds. **Amphipods** are another important type of planktonic crustacean. They have large well-developed eyes at the front of their head to actively seek-out their prey. Some species of amphipods live in association with gelatinous organisms such as jellyfish and salps. **Chaetognaths**, or "arrow worms" belong to a separate phylum, comprising about 120 species. Their name comes from their long, transparent bodies, with side and tail fins. All species are carnivorous with grasping hooks and rows of strong teeth that make them efficient predators on copepods and other small crustaceans. Holoplanktonic **gastropods** (mollusks) swim with their modified foot, which has evolved into two wing-like lobes. The "sea butterflies" (*Limacina* spp.) have calcified shells and feed mainly on phytoplankton. The "sea angel" (*Clione limacina*) has no outer shell and is a specialized predator on *Limacina*.

Appendicularians and **Salps** are holoplanktonic animals with nerve cords and are thus closely related to the vertebrates. They are filter feeders, consuming small food particles, such as phytoplankton and detritus. Appendicularians produce a mucus house with a complicated arrangement of filters to extract particles from the water. The most common species belong to the genera *Oikopleura* and *Fritillaria*. Salps live singly or in colonies and can form massive aggregations of millions of individuals that may play a significant role in marine ecosystems.

Jellyfish are gelatinous animals that belong to the phylum Cnidaria, which are characterized by the possession of nematocysts (stinging cells). Jellyfish are the largest example of plankton and can grow as large as 2 meters wide, with tentacles up to 37 meters. Jellyfish have only primitive organs and nervous systems, and no hard body parts. They are, however, effective predators that catch plankton and larval fish with stinging cells on their tentacles. Common species among **jellyfish** are the moon jellyfish (*Aurelia aurita*), the lion's mane jellyfish (*Cyanea capillata*), and the blue stinging jellyfish (*Cyanea lamarckii*). The colonial **siphonophores** are composed of many specialized individuals which may stretch out up to 50 meters in length like giant fishing nets.

The **comb jellies** resemble the cnidarian jellyfish with their gelatinous bodies, but are members of an unrelated phylum (Ctenophora). The comb jellies use eight rows of ciliary plates for propulsion. These "comb rows" often radiate beautiful color reflections. Ctenophores lack the stinging nematocysts and capture their prey with sticky tentacles (*Pleurobrachia pileus*) or mucous-covered oral lobes (*Bolinopsis infundibulum*).

Jellyfish and comb jellies are ancient animals that have existed in the seas for at least 500 million years. When marine ecosystems become disturbed, jellyfish can proliferate. Human impacts, such as climate change, increased nutrient levels, overfishing, and increased coastal construction, have been cited as contributing to increased frequency of jellyfish blooms. Once an ecosystem has become dominated by jellies, it may become difficult for fish stocks to reestablish themselves, because jellies are predators on fish eggs and larvae.

One example of a large species that recently has increased in deep Norwegian fjords is the deep-water helmet jelly *Periphylla periphylla* (Figure 1.13). It is believed that the increase is due to changes in environmental variables such as salinity, light, and/or variables that could be associated with changes in climate. The preferred depth range of *P. periphylla* is 350–450 m.

Invasive populations of alien jellies can expand rapidly because they often face no predators in the new habitat. Examples include the introduction of the comb jelly *Mnemiopsis leidyi* (Figure 1.14) to the Black Sea in the early 1980s, where it had a catastrophic effect on the entire ecosystem. This alien species has also been introduced via the ballast water of ships to the North Sea and Skagerrak, where it occurs in dense blooms in coastal waters during late summer.

Figure 1.13 The helmet jelly *Periphylla periphylla*. *Source:* E. Svendsen, Norway.

Figure 1.14 The alien ctenophore *Mnemiopsis leidyi* was first observed in the North Sea in 2005. *Source:* Ø. Paulsen, Institute of Marine Research, Norway.

1.4.4.2 Pelagic Fish

Fish species in the pelagic water masses differ somewhat with distance from the coast, but there are common species in both the open ocean and the fjords. Within the pelagic zone, there is a limit to which depth it is practical to sample. In the open ocean, this is usually restricted to maximum 1000 m depth.

Several **epipelagic fish** species that occur from the surface to 200 m depth are common for fjords and the coast, over the continental shelf, and in the open ocean. These include herring (*Clupea harengus*), mackerel (*Scomber scombrus*), blue whiting (*Micromesistius poutassou*). Sprat (*Sprattus sprattus*) is abundant in fjords and over the continental shelf (Zijlstra, 1988) (Figure 1.15). Greater argentine (*Argentina silus*), horse mackerel (*Trachurus trachurus*), and the benthopelagic spiny dogfish (*Squalus acanthias*), lumpsucker (*Cyclopterus lumpus*), and pollack (*Pollachius pollachius*) also occur sporadically within fjords.

Figure 1.15 (a) A typical pelagic fish is the sprat (*Sprattus sprattus*). (b) Pictured are some specimens of the local Lustrafjord population caught in the Fjøsne Bay, September 2016. *Source:* A.G.V. Salvanes.

1.4.4.3 Mesopelagic Organisms

The most numerous fish in the pelagic habitat belong to so-called mesopelagic species (Salvanes and Kristoffersen, 2001). Mesopelagic species are typically found between 150 and 1000 meters deep, often concentrated in one to several deep scattering layers (Sound Scattering Layers; SSLs). Species in these layers undergo diel vertical migration (DVM) to surface waters at night to feed, and stay deep at day time (Marshall, 1971; Salvanes, 2004).

The most common mesopelagic organisms in European deep coastal waters and fjords are pearlside (*Maurolicus muelleri*), northern lanternfish (*Benthosema glaciale*), the shrimps (*Pasiphaea* spp. and *Sergestes* spp.) and krill (*Meganyctiphanes norvegica*) (Figure 1.16). In the open ocean, the

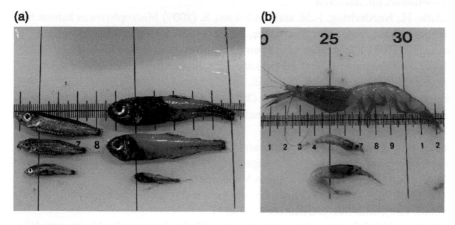

Figure 1.16 Typical mesopelagic organisms from fjord areas. Pictured are specimens caught in Masfjord. (a) Left; *Maurolicus muelleri* and right; *Benthosema glaciale*. (b) Top; *Pasiphaea* spp., middle; *Meganyctiphanes norvegica*, and bottom; *Sergestes* spp. *Source:* A.G.V. Salvanes.

stomatoid fish *Cyclotone* spp. and *Notolepis rissoi* can be very abundant in deep mesopelagic samples. Further information on deep-water species are available in Marshall (1971) and Bergstad (1990a).

1.4.4.4 Deep-pelagic Fish

The bathypelagic zone is below the mesopelagic zone and deeper than 1000 m. The most numerous fish belong to the *Cyclotone* genus (e.g., bristlemouths). Other numerous species include gulper eels, bobtail snipe eels, a few species of macrourids, and brotulids (Marshall, 1971). For further details see Marshall (1971), Merrett and Haedrich (1997), or Randall and Farrell (1997).

References

Aksnes, D.L. (2015) Sverdrup critical depth and the role of water clarity in Norwegian Coastal Water. *ICES Journal of Marine Science*, 72, 2041–2050.

Aksnes, D.L., Aure J., Kaartvedt S. *et al.* (1989) Significance of advection for the carrying capacities of fjord populations. *Marine Ecology Progress Series*, 50, 263–274.

Bergstad, O.A. (1990a) Ecology of the fishes of the Norwegian Deep: distribution and species assemblages. *Netherlands Journal of Sea Research*, 25(1), 237–266. DOI:10.1016/0077-7579(90)90025-C

Bergstad, O.A. (1990b) Distribution, population structure, growth and reproduction of the roundnose grenadier *Coryphaenoides rupestris* (Pisces: Macrouridae) in the deep waters of the Skagerrak. *Marine Biology*, 107(1), 25–39. DOI:10.1007/BF01313239

Bergstad O.A., Bjelland O. and Gordon J.D. (1999) Fish communities on the slope of the eastern Norwegian Sea. *Sarsia*, 84(1), 67–78. DOI:10.1080/00364827.1999.10420452

Bjelland, O. and Holst, J. (2004) Other fish species and fish communities, in *The Norwegian Sea Ecosystem* (ed. H.R. Skjoldal), Tapir Academic Press, Trondheim, pp. 357–370.

Christie, H., Norderhaug, K.M. and Fredriksen, S. (2009) Macrophytes as habitat for fauna. *Marine Ecology Progress Series*, 396, 221–233. DOI:10.3354/meps08351

Fosså, J.H. (1991) The ecology of the two-spot goby (*Gobiusculus flavescens* Fabricius): the potential for cod enhancement. *Proceedings of the ICES Marine Science Symposia, 1991.* pp. 147 –155.

Heessen, H.J.L., Daan, N. and Ellis J.R. (2015) *Fish Atlas of the Celtic Sea, North Sea, and Baltic Sea: Based on International Research-vessel Surveys*, Wageningen Academic Publishers, Wageningen.

Hjort, J. and Gran, H. (1899) Currents and pelagic life in the Northern Ocean, in *Report on Norwegian Marine Investigations 1895–97.* vol. VI (eds. J. Hjort, H. Gran and O. Nordgaard), Bergens Museums Skrifter, Bergen.

Lewis, J.R. (1961) The Littoral Zone on rocky shores: A biological or physical entity? *Oikos*, 12(2), 280–301. DOI:10.2307/3564701

Mann K.H. (2000) *Ecology of Coastal Waters: With Implications for Management, 2nd edition*, 2nd edition ed.: John Wiley & Sons.

Mann K.H. and Lazier J.R. (2006) *Dynamics of Marine Ecosystems: Biological-physical Interactions in the Oceans*, 3rd edition ed.: Blackwell Publishing.

Marshall, N.B. (1971) *Explorations in the Life of Fishes*, Harvard University Press, Cambridge.

Melle, W., Ellertsen, B. and Skjoldal, H. (2004) Zooplankton: the link to higher trophic levels, in *The Norwegian Sea Ecosystem* (ed. H.R. Skjoldal), Tapir Academic Press, Trondheim, pp. 137–202.

Merrett, N.R. and Haedrich, R.L. (1997) *Deep-sea Demersal Fish and Fisheries*, Chapman & Hall, London.

Moen, F.E. and Svensen, E. (2004) *Marine Fish & Invertebrates of Northern Europe: AquaPress*, Southend-On-Sea.

Omand, M. M. and Mahadevan, A. (2015) The shape of the oceanic nitracline, *Biogeosciences*, 12, 3273–3287, DOI:10.5194/bg-12-3273-2015

Postma, H. and Zijlstra, J.J. (1988) *Continental Shelves (Ecosystems of the World)*, Elsevier Science Ltd., Amsterdam.

Randall, D. and Farrell, A.P. (1997) *Deep-sea Fishes*, Academic Press, San Diego.

Rex, M.A. and Etter, R.J. (2010) *Deep-Sea Biodiversity: Pattern and Scale*, Harvard University Press, Cambridge.

Røstad, A., Kaartvedt, S. and Aksnes, D.L. (2016) Light comfort zones of mesopelagic acoustic scattering layers in two contrasting optical environments. *Deep Sea Research Part I: Oceanographic Research Papers*, 113, 1–6.

Sætre, R. (2007) *The Norwegian Coastal Current – Oceanography and Climate*. Tapir Academic Press, Trondheim.

Salvanes, A.G.V. (2004) Mesopelagic fish, in *The Norwegian Sea Ecosystem* (ed. H.R. Skjoldal), Tapir Academic Press, Trondheim, pp. 301–314.

Salvanes, A.G.V. and Kristoffersen, J.B. (2001) Mesopelagic fish (life histories, behaviour, adaptation), in *Encyclopedia of Ocean Sciences* (eds: J. Steele, S. Thorpe and K. Turekian), Academic Press Ltd., San Diego, pp. 1711–1717.

Salvanes, A.G.V. and Nordeide, J.T. (1993) Dominating sublittoral fish species in a west Norwegian fjord and their trophic links to cod (*Gadus morhua* L.). *Sarsia*, 78(3–4), 221–234. DOI:10.1080/00364827.1993.10413536

Skjoldal, H.R. (2004) *The Norwegian Sea Ecosystem*, Tapir Academic Press, Trondheim.

Staby, A., Røstad, A. and Kaartvedt, S. (2011) Long-term acoustical observations of the mesopelagic fish *Maurolicus muelleri* reveal novel and varied vertical migration patterns. *Marine Ecology Progress Series*, 441, 241–255. DOI:10.3354/meps09363

Stedmon, C.A. and Markager, S. (2003) Behaviour of the optical properties of coloured dissolved organic matter under conservative mixing. *Estuarine Coastal and Shelf Science*, 57, 973–979.

Wheeler, A.C. (1969) *The Fishes of the British Isles and North-West Europe*, Macmillan, London.

Zijlstra, J.J. (1988) The North Sea ecosystem, in *Continental Shelves (Ecosystems of the World)* (eds. H. Postma and J.J. Zijlstra), Elsevier Science Ltd., Amsterdam, pp. 231–278.

2

Planning Marine Field Studies

Jennifer Devine*, Keno Ferter, Henrik Glenner, Jon Thomassen Hestetun, Knut Helge Jensen, Leif Nøttestad, Michael Pennington, David John Rees, Anne Gro Vea Salvanes, Kjersti Sjøtun and Arved Staby

The first step when designing a scientific field survey is to define the objectives of the field study, for example an objective could be to estimate the age and length distribution of a fish population. The next step is to determine the best equipment and methods for sampling the target population, for example a trawl that will effectively and efficiently capture a targeted fish species in the field. By standardizing sampling gear and field sampling procedures the samples will represent accurately and consistently the conditions at every station. Thus, for example, an index of abundance based on data from all the stations of a well-designed and executed survey can reliably track changes in abundance. The last step is to design an appropriate and statistically sound survey design for sampling the population.

This chapter begins by presenting various ways to design a scientific survey for field studies, including a description of the most used designs. This is followed by descriptions of some currently employed survey designs for littoral and benthos field studies: an oceanic survey to estimate the abundance of mackerel; a bottom trawl survey to monitor demersal fish populations; a hydroacoustic survey to study the diel vertical migration of mesopelagic organisms, and lastly, a survey design to study barotrauma in physoclistous fish species.

2.1   Survey and Sampling Design

A **survey** is a method of collecting information using observations to gather information from a fraction of the individuals in a population to say something about the whole population. The word population has a very broad meaning when it comes to surveys. The process of picking out single objects, items, or organisms on which to take the measurement(s) is called **sampling** and the collection of measurements (the sample) should reflect or represent the population under study. In other words,

---

* Lead author; co-authors in alphabetical order.

Marine Ecological Field Methods: A Guide for Marine Biologists and Fisheries Scientists,
First Edition. Edited by Anne Gro Vea Salvanes, Jennifer Devine, Knut Helge Jensen,
Jon Thomassen Hestetun, Kjersti Sjøtun and Henrik Glenner.

the sample should be a small version of the studied population and must contain all the characteristics (and their variability) inherent in the larger population. Picking out organisms randomly from the total population will intuitively produce a sample with some of the same properties as the total population, but a larger sample (more observations) is likely to be closer to the total population.

Every field investigation aiming to characterize a population must be based on a specific **survey design** in order to produce unbiased estimates of the wanted population traits along with the corresponding estimates of uncertainty or level of precision. The design of a survey can influence the outcomes of the analysis, therefore, implementing a design that will allow for answering the question without **bias** (i.e., a systematic favoritism that gives misleading results or results that differ from the true characteristics of the population) is crucial. When deciding on a survey design, one must have the aims of the survey in mind: is the purpose of the survey to determine, for example, the relative abundance of a particular species; patterns in behavior, such as diel vertical migration; patterns in species distribution; transport of eggs and larvae with currents; change in species composition due to environmental factors or a disrupting event. Other questions to consider include: how often will the survey be done, that is, is it a one-time event or is it long-term monitoring? If it is long-term monitoring, could the needs of the survey change in time (i.e., should one design the survey to be adaptable for alternate needs)? Are there large movements of the species/population of interest within the survey area?

Designing a survey should follow the following steps:

1) **Determine the objectives of the survey.** If the objective is to map species distributions across one or more habitats, the design will be different from that used to estimate (with precision) the abundance of key stocks for assessment and management purposes.
2) **Determine the populations(s) to be sampled.** The survey area should include all areas where the population is thought to occur and the boundaries should extend slightly beyond the population boundary; this is to ensure total coverage of the population. This should also determine whether temporal fluctuations need to be taken into account, that is, are surveys needed at several times throughout the year.
3) **Determine how the sampling sites will be chosen (sampling design).**
4) **Determine the data to be collected; this includes *what* to measure and *how many* measurements to take.** One should carefully determine how much data is needed. A survey is costly; there is a trade-off between the amount of data that can be collected and the time (and thus cost) incurred by collecting that amount of data. However, one must never collect too few as there is usually no opportunity to go back and collect more data.
5) **Determine the amount of precision needed.**

Perform a trial and make modifications if the objectives are not fully met. Modifications to the survey design at the beginning of sampling may increase precision and cost less than modifications mid-way through a sampling program. During a survey, additional data should be collected to ensure the survey is still obtaining its objectives (experimentation and quality assurance). For example, areas outside of the defined survey area might occasionally be sampled to see if species distribution has changed and thus the survey area should be modified.

2.1.1 Survey Design

Michael Pennington, Institute of Marine Research, Norway

There are two standard survey designs for field studies. One common design is a **stratified random survey**. For these surveys the area to be covered is divided into subareas (**strata**) and within each stratum the sampling sites (**stations**) are chosen randomly. At each station, samples are collected using standardized gears and field sampling procedures. The geographical stratification of a survey area is chosen either because it is believed that the targeted populations are more similar within a stratum than in the entire population and if so, the resulting estimates will be more precise; or for management purposes, for example if abundance estimates are needed for particular subareas. An advantage of a stratified design with randomized stations within strata is that the resulting estimates are theoretically unbiased, that is as sample size increases, the estimates converge to the true value. Another advantage is that the true variance of the estimate is easy to calculate. In practice, because samples taken at stations close together tend to be similar, stations are often selected using a **buffered sampling** technique, which sets a minimum distance between the randomly selected stations (Carlson, Folmer, Kanneworff, *et al.*, 2000).

Sampling effort can be distributed among the strata in many ways. Two common sampling strategies are: allocate sampling intensity based on the variability of the population in each stratum, which theoretically will give the most precise estimate for the entire area; the other is to sample, for example, important management areas more intensely. Usually the safest choice, if feasible, is to have the sampling effort in each stratum proportional to its size, that is, the sampling intensity is uniform over the entire survey region.

The other common survey design for field studies is a **systematic survey**. The most used design is to overlay a grid on the survey area and use the vertices to locate the sampling stations. A systematic survey can be combined with a stratified design by either overlaying a stratification over the grid, again based on subareas that are thought to be similar, or choosing a different grid for each stratum or a subset of strata. The single grid design will result in uniform sampling coverage throughout the entire survey area, while the second strategy is used to vary sampling effort in selected subareas.

A systematic survey can produce more precise estimates than a random survey if there is positive correlation of factors, such as abundance or species composition, between stations that increases as the sampling stations get closer together. For marine surveys, the correlation between stations is often large enough that a systematic survey will generate more precise estimates than a random survey. As mentioned, it is straightforward to estimate the variance for random surveys, but this is not the case for systematic surveys. There are several suggested ways to estimate the variance for systematic surveys, either model-based or by grouping stations, but there is no certainty whether the estimates are accurate (Cochran, 1977). However, in some situations it may be better to have a more precise estimate of some quantity with an associated biased or unknown variance estimate, than, for example, a random survey producing a less precise estimate with an unbiased variance estimate.

There are, of course, other survey designs for marine field surveys other than stratified random surveys or systematic surveys. For example, some surveys keep a subset of stations fixed and the other stations are chosen at random. One reason for keeping some stations fixed is that yearly changes can sometime be detected and measured more accurately using fixed stations than by using random stations (Warren, 1994). A **two-phase adaptive** design allocates sampling within a stratified design in two stages; the first stage is used to determine which strata, based on high variance, should receive additional samples during the second stage (see Francis, 1984; Jolly and Hampton, 1990; Salehi *et al.*, 2010 and references therein). This is used when prior information on the distribution of the population is poorly known or where the density of the population is likely to vary within defined strata, but the extent of the variability is not known, that is, high density areas cannot be known in advance. For more details on the design and analysis of surveys, see Kish (1965), Cochran (1977), and Gunderson (1993).

2.1.2 Sampling at a Station

Michael Pennington, Institute of Marine Research, Norway

An important aspect for designing a field survey is to determine how and how much to sample at each station. Fish (or flora and fauna in general) sampled during a field survey are not a random sample of individual fish from the entire population, but a sample of n clusters, one cluster from each station. Since fish caught together are usually more similar than those in the general population, a total of m fish collected from n clusters will contain less information about the distribution of the variable of interest for the entire population than if m fish were randomly sampled from the population – which is impossible to do in practice.

One measure of the amount information in a sample from a complex sampling scheme, such as marine field surveys, is the **effective sample size**, which is defined as the number of individuals that would need to be sampled at random so that the estimates generated by simple random sampling would have had the same precision as the estimates obtained based on the more complex sampling scheme (Kish, 1965; Skinner, Holt and Smith, 1989). In particular, the effective sample size is a transparent and efficient way to measure the amount of information for estimating, say, mean age or length contained in a cluster sample from marine surveys. Unfortunately, because of positive intracluster (**intrahaul**) correlation, the effective sample size is generally much smaller than the number of fish sampled (Pennington and Vølstad, 1994, Pennington, Burmeister, and Hjellvik, 2002; Nelson, 2014). A typical example is from the Norwegian winter Barents Sea survey. Over a 14-year period, an average 149 933 Northeast Artic cod were sampled and measured each year, while the average effective sample size was only 86 fish per year (Pennington, Burmeister, and Hjellvik, 2002).

The precision of estimates of other population characteristics, such as age distributions, can also be relatively low compared with the number of fish sampled if an attribute or measurement is more similar for fish caught together than for those in the general population. For example, the precision of estimates of mean stomach contents (Bogstad, Pennington, and Vølstad, 1995), or diet composition (Tirasin and Jørgensen, 1999), or estimating sex ratios for a fish population (Francis, 2014), etc. can be relatively low because of positive intrahaul correlation.

A relatively small effective sample size implies that it is best to take small samples at as many locations as possible, which is the only way to increase the effective sample sizes for estimating population attributes. For example, it has been shown that trawl or dredge tows of short duration are in general more efficient for estimating stock abundance and population characteristics than long tows (Godø, Pennington, and Vølstad, 1990; Pennington and Vølstad, 1991; Gunderson, 1993; Carlsson *et al.*, 2000). Therefore, one way to collect samples from more locations and improve overall survey efficiency without increasing survey cost is to reduce tow duration and use the time saved to increase the number of survey stations (Pennington and Vølstad, 1994).

Many methods exist for choosing the sampling location and only the most used are mentioned here. If sampling locations change for each sampling event (i.e., each survey) and their placement within a strata is randomly chosen (selected without prior knowledge and all sampling locations are considered possible), then the survey has a **random** sampling design. Sampling locations may also be **fixed**, the same locations are used for every survey, **systematic**, **quasi-random** (sampling appears random, but does not comply with the true definition of random), or a **combination** of random and fixed. Regardless of which type of sampling placement is being used, samples from each location must be independent.

Random station placement is easy to do and yields unbiased estimates with known error variance. The disadvantage is that it is inefficient when populations are spatially autocorrelated; stations placed close to each other will replicate information and will not provide independent samples. In this case, the statistical population is not the same as the biological population and instead is the (theoretical, but valid) population of all possible catches.

Systematic sampling involves sampling according to a predetermined pattern, such as 10 m spacing, where the location of the first station is assigned randomly. Systematic sampling is often used when assigning transect placement, where the location of the first transect is randomly chosen. Using a systematic design allows for easier planning of the survey and ensures station density is the same throughout the survey area. The disadvantage of this method is that the sampling variance cannot be estimated without bias because stations are not independent.

Fixed stations have the advantage of being directly comparable over time. Such placement has been shown to be superior to random in certain situations, such as monitoring the effect of disturbance if the disturbance zone is well defined and there is no lessening of effects with distance from the disturbance site (e.g., localized physical disturbance, disturbance along a chemical gradient; Morehead, Montagna, and Kennicutt, 2008). Fixed stations have often been used when investigating trends in abundance of populations over time, but have the disadvantage that if species' distributions change between sampling events, bias will occur.

If **quasi-random**, stations are selected at random from a list of known (previously used) stations. Placement is not truly random because they have been used during other surveys, but, on a survey-to-survey basis, which stations are used will vary randomly. This is often done in cases where there is limited area to place stations. Use of this placement may have repercussions on the statistical power of the survey design.

Combination placement uses a combination of fixed stations and randomly-decided stations. This type of design is often useful when sampling in areas with limited sampling availability for some habitats, stratum, or grids. A combination sampling design is often used to assess the performance of a fixed-station placement relative to a random placement in estimating annual changes in biomass.

How samples are allocated between strata can be by the following schemes:

- **Equal**: if strata are of approximately equal size, the number of samples can be split equally between the strata.
- **Proportional**: samples are allocated according to size of the strata, i.e., more samples are allocated to larger strata.
- **Optimal**: the allocation of samples is such that the variance of the mean (and total) is minimized. Stratum that have higher variances or a larger proportion of the population will require a higher sampling intensity. Areas with higher abundance tend to have higher variability; the greater the variability, the more samples needed to obtain the same precision as a strata with low variability. This allocation requires prior knowledge of the abundance or variability between strata or, at least, a good guess of the expected size of the variability. If the variability for all strata is equal, the optimal sampling scheme will be a proportional sampling scheme.

The number of stations to sample within each stratum is determined by specific formulas, which depend on the allocation scheme (see bullet list above).

2.2 Littoral Survey Design

To design a study in the littoral zone, a clear view of the goals of the study and the questions one wants to answer is needed. In some cases, the aim will be to make a descriptive study of one or several littoral localities, which can be performed with different degrees of resolution or thoroughness. In other cases, the aim will be to measure potential changes in the shore environment due to industry or other anthropogenic activity. Finally, the aim can be basic research, with the objective to test hypotheses on interaction, patterns, or processes within littoral communities. This will often involve setting up experiments, manipulating communities at a number of localities, and then measuring the results.

A **baseline study** is a descriptive one-time study of one or more survey localities. Typically, a baseline study is done before the initiation of some kind of activity that may potentially be harmful to the environment. Qualitative shore surveys may be used to quickly describe the dominating species and zonation pattern of the littoral zone in a wide area. However, often a more thorough approach is required, involving a large number of survey sites and a high precision level of data sampling. These more time-consuming methods provide more data and thus allow a more complete description of the site(s). As is always the case, using a more time-consuming method means less area can be covered within the same time frame.

For environmental surveys, the aim of the study is to measure whether there is any response in the nearby littoral community to a certain planned future event. Typically, this can be the establishment of industry, an aquaculture farm, a sewage treatment plant, or large-scale construction work in a shore or nearshore area. This kind of study design is referred to as an **impact study**. An impact study should be planned as a BACI study (Before, After, Control, Impact) and is a study where a number of littoral stations close to the impact site and control sites further away are surveyed both before and after the impact (Smith, 2002). Surveying the sites before and after impact allows for the measurement of any changes, while control stations are used to control for natural fluctuations in community structure. This kind of study may be expanded in various ways, for instance, in the number of sites and control localities chosen.

A before–after approach may not be sufficient to look for long-term effects of a persistent suspected impact source. Thus for nearshore areas with anthropogenic influence (e.g. sewage, industry, aquaculture), monitoring programs are set up to revisit the same study localities at set intervals throughout the operation of the facility. Typically, an environmental monitoring program with regular surveys will accompany the industry or anthropogenic activity. While the focus is on the littoral zone here, environmental monitoring programs, in most cases, involve sampling a combination of littoral localities, benthic stations, and the water column, providing a comprehensive coverage of the nearby marine environment.

Testing specific hypotheses about species interactions or community patterns and persistence is necessary to explain spatial and temporal changes. Testing hypotheses is normally done by designing field experiments. Such experiments can, for example, be to remove specific organisms to examine trophic interactions, or to remove all organisms to study resettlement and processes during regrowth of a community. The outcome of some classic experiments have demonstrated, for example, how zonation patterns are formed, or how predation can influence the community structure (e.g. Schonbeck and Norton, 1980; Connell, 1985; Paine, 1974). Increasing the knowledge of the interactions of the littoral communities allows for a better interpretation of general changes in littoral localities.

Biological data: Which parameters to sample depends on the aim of the study, the methodology, and on the amount of time and funds available. It is common to include population parameters such as number of species, biomass, abundance, percentage cover, density, number of individuals, or individual size distribution. In many cases, more specific measures are used, such as data on certain dominating or indicator species, or measurements of specific chemical parameters from organism tissue.

In addition to collecting data on the biota, data on the physical environment of the sites should be included because physical factors will also influence the community composition. By analyzing both biological and physical data together, the factors that are most important in shaping community structure can be determined.

Physical data: Physical parameters are important in explaining biological variation, and depending on the study it is common to record information about aspect, degree of inclination, topographical variation, and wave-exposure of the study sites. Using a leveling instrument and measurement stick, inclination as well as the height above water level of zones of dominating organisms can be measured. Measurements of temperature, salinity, light, and nutrients will show strong temporal variation at a site. Temperature and salinity is relatively easy to monitor by loggers over a predetermined time period.

2.2.1 Sampling Methods

A number of sampling methods exist based on the precision level and type of data required. Vertical transects can be used to map different zones of dominating organisms at the study localities. Vertical transects can be made by measuring the height of dominating organisms in relation to water level and Chart Datum. A transect can be extended into the subtidal part of the littoral zone through scuba diving or by use of a ROV (Remotely Operated Vehicle). For large-scale surveys, remote sensing using airborne near-infrared equipment or echogram soundings can be used in conjunction with GIS software to map large areas.

When a high precision level is needed, detailed community analyses are necessary. This can be achieved using sampling plots or quadrats and making a detailed recording of organisms present inside each sample plot. Abundance data can be collected in different ways.

Quadrat analysis, where a number of plots (quadrats) will represent subsamples of the community at the examined location, is one method of field sampling. The subsamples should be representative of the entire community; this means that the spatial distribution of the plots at the locality is an important consideration. In the intertidal, the physical gradient of air exposure is very important in determining the community composition. Therefore, if one is to sample the whole community, the sample squares should be distributed accordingly and cover both the upper, middle, and lower intertidal. In addition, the possibility of autocorrelation of species abundance between the sample plots must be taken into account when deciding on plot location. When assigning replicates, this autocorrelation must be taken into account in the study design.

2.3 Benthos Survey Design

A benthic survey can take many forms depending on the scope of the survey, the general purpose, the seafloor type, what part of the benthic fauna is investigated, and the equipment available. Surveys may vary from large scientific cruises using a variety of towed gear, ROV and corers, to simple nearshore surveys using grab sampling and smaller towed equipment. The traditional tools are towed gear, in the form of a variety of bottom trawls, sleds, and dredges, as well as grabs and corers. In recent times, ROVs have become more much more prominent, allowing high definition *in situ* imaging of live organisms and targeted sampling of biota of interest. Landers and baited traps are also used to capture specific organisms.

The sheer vastness of the ocean seafloor means that much of the benthic sampling is still exploratory in nature; simply trying to map the large-scale patterns of biotic communities is a huge undertaking in itself, given the difficulty of sampling, together with the huge biodiversity of benthic invertebrates. The seafloor in most of the world's oceans remains severely under sampled. Thus, most benthic gear provides qualitative or, at best, semi-quantitative rather than quantitative data. This means that certain types of information, such as biomass or density as a function of seafloor area, cannot be precisely calculated. Estimates are also compounded by the patchiness of benthic communities, which is influenced by topography, bottom substrate, currents, and other, less well-known variables.

Therefore, the most common type of benthic surveying remains a qualitative description of a certain area, often as part of a project comprising a cruise or series of cruises, and including other types of oceanographic and marine biological data, such as pelagic CTD (conductivity, temperature, depth) and biological information, as well as geological and microbiological components. A series of benthic stations can be designated using different types of equipment, either following specific patterns or around topographic points of interest, such as a vent system or seamount. Sampling can be built around specific hypotheses regarding community processes or relationships using any of a number of methods.

The type of towed equipment used depends on the substrate. Different bottom trawls and sleds are designed to either skip over or dip into the sediment for soft-bottom substrates, while sturdier dredges are more common for hard-bottom habitats. The type of sea bottom substrate can often be guessed by looking at the topography of the sampling location. Using equipment on the wrong type of substrate can result in few samples or, in certain cases, damage or loss of equipment.

ROV sampling has become common in benthic surveys. ROVs provide information on *in situ* characteristics and behavior of live specimens. A common sampling design using ROVs includes video transects, with the objective to map larger biota (e.g., biogenic habitat such as large coral reefs and the organisms living in and on the reefs). Biota of interest can be retrieved depending on the equipment mounted on the ROV.

In certain cases, such as ecological surveys, baseline studies, or environmental monitoring, quantitative data is desired. Quantitative data can be extrapolated, given a number of caveats, to larger areas. Grab and corer sampling allows quantitative sampling because the volume of sediment can be measured; however, these sampling gears are not useful for obtaining quantitative data from hard-bottom substrate. For hard substrate habitats, video mosaic analysis is useful, but limited in the amount of biota that is identifiable from pictures.

The sampling methods presented in Chapter 3 apply to benthos living in the sub-littoral zone along the foreshore, out to the continental shelf and down to abyssal depths. Specifically, this means the part of the seabed where it is not possible for a human to reach the bottom without technical assistance. This designation is not very descriptive and includes a number of different sediment types and habitats. Which habitat is dominating in a certain area depends on geological history, hydrographical factors, water chemistry, temperature, and species composition. For example, soft sediments can be homogeneous for thousands of kilometers, but are often highly heterogeneous and change dramatically within a few square meters. If the assignment of a marine biologist is to study or describe such habitat, detailed information about the topology and composition of the seafloor is required. That would definitely be requested in studies of any terrestrial system. No study of, let's say the grasshopper fauna of a well-demarcated terrestrial area, would be taken seriously if it were unknown whether the sampled animals came from a field or a meadow! This is, however, the situation when sampling marine soft-bottom sediments unless submarines, ROVs, or divers are involved in the research; options that in most cases are either practically impossible or would add tremendous cost to the research. Benthic sampling sites are often located hundreds of meters below the deck of the research vessel, from which one, in complete blindness, tries to operate crude sampling devices with a metal wire at the speed and direction of the research vessel. The organisms one eventually gets on the deck might come from an area the size of a soccer field, from habitats as different from each other as a cornfield is from a watery meadow. From the organisms collected from such a haul one will, without more than a restricted prior knowledge about the seafloor, often have to come up with a reasonable interpretation about the biology of the sampled organisms, their habitats, and the community structure. From the view of a terrestrial biologist, this would be close to unmanageable working conditions, but this is the sort of setting in which marine biological researchers studying soft-bottom sediments are working. The quality of the data is not comparable to data from terrestrial environments, but they are better than nothing; marine biologists just have to be modest in the requirements for data quality and do the best with what we have.

2.3.1 Mapping the Biodiversity of Sognefjord – An Example of a Multi-sampling Approach

David John Rees, University of Bergen

In 2011 the University of Bergen, the Institute of Marine Research (IMR), the University Museum of Bergen, Umeå University and the Norwegian Institute for Water research (NIVA) embarked on a joint project, financed by the Norwegian Biodiversity Information Centre, aiming to produce a comprehensive inventory of the benthic fauna in Sognefjord and to map species distribution. This work was intended as a foundation for diverse subsequent investigations that include biodiversity patterns, population structure, and genetic variability.

Sognefjord is the longest and deepest fjord in Norway, stretching over 200 km inland with numerous side-fjords branching off from the central fjord. Comprehensive physical sampling is a complex and challenging operation considering working depths in excess of 1200 m. Our goals were to collect benthic samples from throughout this fjord network, including fjord branches and the oceanic area lying outside the mouth of the fjord itself, and to describe and map the benthic nature types represented in the fjord system.

Mapping benthic fauna collected blindly by classical sampling methods is insufficient to provide a clear picture of the environmental setting of the different faunal communities and to identify the nature types present in the fjord. Therefore, targeted video surveys and ROV deployments were used to complement classical sampling methods, ultimately enabling description, characterization, and mapping of the benthic fauna nature types as well as generation of species lists for the region.

Benthic sampling in the central fjord basin involved deploying gear to depths of over 1200 m as well as gear deployment in shallower side-fjords (100–800 m). To have the best chance of achieving our aims we employed a multi-sampling approach with diverse "classical" sampling gear and also state-of-the-art ROV and video sampling equipment. In addition, we incorporated data from several previous sampling efforts in Sognefjord, dating back to the 1940s.

2.3.1.1 The Objectives of the Project

1) To obtain a complete faunistic overview of the macro- and megabenthic fauna and categorize marine benthic nature types of Sognefjord using benthic sampling and video.
2) To identify species new for the area, for Norway, and to science.
3) To define benthic nature types that are characteristic for West Norwegian sill fjords, which should be useful in future mapping efforts of other Norwegian fjords.
4) To recruit and train young researchers in the field of biosystematics, combining morphological and molecular data in taxonomic diagnostics.
5) To compare differences in fauna distributions and nature types in the different parts of the fjord.

2.3.1.2 Sampling Strategy and Sampling Design

To describe the biodiversity of benthic organisms of an area as enormous as Sognefjord is an enormous task. Due to economic and time limitations, a complete coverage of the benthic fauna in the fjord is not possible. To obtain the best possible estimation of biodiversity, the sampling strategy had to be optimized in such a way that all the habitats of the fjord were represented. We applied complementary sampling and surveillance suited to collecting epi/infauna, on hard- and soft-bottoms, or for large mobile/small sedentary animals. In general, ROV-based video observations were used as a quantification tool of epibenthos (Buhl-Mortensen *et al.*, 2015) while grabs, sledges, dredges and bottom trawls provided qualitative data.

2.3.1.3 Methods and Sampling Activities

Five classical sampling cruises were conducted between 2011 and 2016, together with four video/ROV cruises. Our sampling primarily focused on the under-studied deeper parts of the central fjord but also involved a large number of stations in side-fjords and, as such, represented a wide range of potential benthic habitats. A network of sampling stations was established involving grabs (Van Veen grab), sleds (Rothlisberg-Pearcy epibenthic sled (RP), Beyer sled), trawl (shrimp trawl, Agassiz trawl), triangular dredges, and Isaac Kidd (MIK) nets for mesopelagic plankton sampling. The diverse sampling gear employed allowed recovery of organisms from a wide range of bottom substrates and complemented previous more tightly focused historical sampling efforts such as the grab sampling undertaken by SAM-Marin (1988–2007). Over 120 gear deployments were made resulting in collection of over 20 000 physical specimens (e.g. Figure 2.1).

(a)   **(b)**

Figure 2.1 **Still images from a video equipped ROV of two typical invertebrate inhabitants of a west-Norwegian fjord. (a) The squat lobster *Munida sarsi* in a soft bottom habitat. (b) The sea star *Hippasteria phrygiana* on the rocky substrate of an undersea mountain.** *Source:* H. Glenner.

2.4 Oceanic Survey Design

Surveys in the open ocean can be for many purposes. Here we will focus only on fisheries surveys and we will illustrate the issues underlying some of the survey designs using case studies.

2.4.1 Pelagic Trawl Survey for Abundance Estimation of Mackerel

Leif Nøttestad, Institute of Marine Research, Norway

2.4.1.1 Background

Northeast Atlantic (NEA) mackerel (*Scomber scombrus*; Figure 2.2), is a widely distributed, highly migratory pelagic fish (Trenkel, Huse and MacKenzie *et al.*, 2014). Mackerel play a key ecological role in oceanic and coastal ecosystems and currently support one of the most valuable commercial fisheries in the North Atlantic (Trenkel, Huse and MacKenzie *et al.*, 2014). In recent years, a large fishery targeted mackerel in the Northeast Atlantic. The geographic range of the mackerel fishery has recently expanded, creating issues with the spatial distribution of the stock, resulting in the highly uncertain stock assessment as determined by the International Council for the Exploration of the Seas (ICES). ICES is an organization that develops science and advice for Northeast Atlantic commercial species (www.ices.dk). Limited tuning data (e.g., surveys) have created challenges for the assessment and management of NEA mackerel. ICES has repeatedly stated the need for an annual age-disaggregated abundance index of this stock (ICES, 2014). These were the motivations for the establishment of an international pelagic trawl survey in 2007 (Nøttestad *et al.*, 2016b).

2.4.1.2 Primary Objectives

The main objectives were to quantify the changes in distribution, abundance, and density of mackerel that feed in the Nordic seas using swept-area trawl survey methods and to estimate the precision of these estimates. We aimed to develop annual age-disaggregated abundance indices, along with associated precision estimates, and to evaluate their applicability to be used as a tuning series in the stock assessment.

Figure 2.2 Northeast Atlantic mackerel (*Scomber scombrus*) swimming inside a pelagic trawl. *Source:* L. Nøttestad.

2.4.1.3 Survey Design

The annual survey uses a stratified systematic design with one Primary Sampling Unit (PSU) per stratum (Figure 2.3). The design resulted in some analytical challenges in providing unbiased estimates of abundance and their associated variances.

Parallel east–west transects were used. Fixed sampling stations were taken at pre-determined geographical positions within strata. Standardized protocol was used, that is, all vessels used the same gear, rigging, towing, and sampling procedures. Trawling protocol included using a towing speed of 5 knots and tow duration of 30 minutes for all vessels and nations from 2007 to 2016 (ICES, 2013c, 2016).

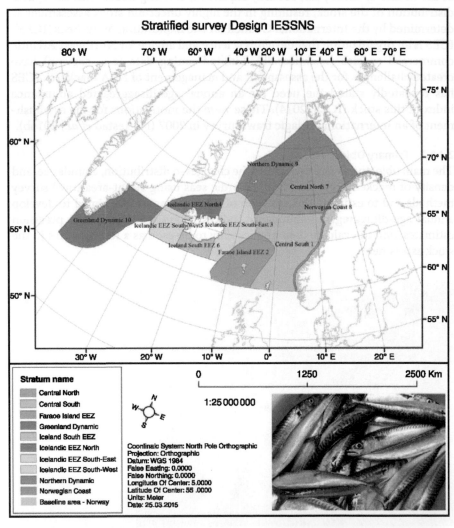

Figure 2.3 Map showing the ten strata used for estimation of mackerel biomass during the International Ecosystem Summer Survey in the Nordic Seas (IESSNS). *Source:* L. Nøttestad.

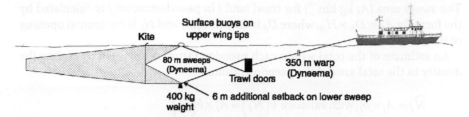

Figure 2.4 Side view of the Multpelt 832 as deployed from a vessel. The Multpelt 832 is rigged with 350 m of warp, 80 m of sweeps, buoys, a kite, and 400 kg chain weights on each lower wing. *Source:* R. Jakobsen.

2.4.1.4 Equipment

Research vessels and chartered commercial fishing vessels from Norway, Faroe Islands, and Iceland were used for conducting the mackerel monitoring survey in the Nordic seas and adjacent waters during July–August from 2007–2014. A Multpelt 832 pelagic sampling trawl was used on all vessels (ICES, 2013c). The rigging is strictly standardized (Figure 2.4) and is detailed in ICES (2015b, 2016).

2.4.1.5 Sampling

Trawls were towed in the surface waters. Flotation was assisted by a kite attached to the top of the net and floats attached to the wings and the headline. The trawl tow track was in a curved "banana" shaped pattern to keep the trawl outside of the wake of the ship. Towing and sampling were done over a 24-hour period (the entire day/night). Details on fish and sampling protocol are given in ICES (2015b, 2016).

2.4.1.6 Assumptions

The main assumption was that all mackerel inside the width of the trawl opening (~65 m) were caught. All mackerel in the vertical dimension were assumed captured, where no mackerel were assumed to be swimming below 30–35 m depth (see Nøttestad *et al.*, 2016a). No diurnal differences in capture were assumed to exist, hence towing at all times of day (24 hours). Samples between pseudostrata were assumed to be independent.

2.4.1.7 Computations

Before estimating total biomass of mackerel for all strata, biomass (with variance) must be calculated from the trawl samples within each stratum. At each trawl station i within pseudostratum j, the number of fish caught N_{ji} is divided by the area swept by the trawl A_{ji} to yield a standardized observation of abundance by square km for the PSU (trawl haul):

$$p_{ji} \frac{N_{ji}}{A_{ji}} \tag{2.1}$$

The swept area (A; kg km^{-2}) for trawl haul i in pseudostratum j is calculated by the formula $A_{ji} = D_{ji} \times H_{ji}$, where D_{ji} is distance (m) and H_{ji} is horizontal opening of the trawl (m).

An estimate of the total within each pseudostratum is obtained by scaling the density to the total area of the pseudostratum:

$$\widehat{N}_j = A_j \times \hat{\rho}_j, \text{with variance } \hat{v}\left(\widehat{N}_J\right) = A_j^2 \times \hat{v}\left(\hat{\rho}_j\right).$$

The total abundance over all pseudostrata is $\widehat{N} = \sum_{j=1}^{J} \widehat{N}_j$, with variance $\hat{v}(\widehat{N}) = \sum_{j=1}^{J} \hat{v}\widehat{N}_j$, provided independent sampling among pseudostrata.

2.4.1.8 Results

The number of original strata in the Nordic seas containing mackerel increased extensively from 2007 to 2014, suggesting a geographical expansion from ~1.3 million km^2 in 2007 to ~2.9 million km^2 in 2014. In 2013, mackerel had expanded north to Svalbard (Berge *et al.*, 2015). The mackerel were distributed along the northern part of the Norwegian coast and west into east Icelandic waters from 2007. By 2012, mackerel were distributed into southeast Greenland waters. Calculations of total biomass indices from the swept area showed an increase from 1.96 million tonnes (RSE = 30%) in 2007 to 8.77 million tonnes (RSE = 8%) in 2014.

2.4.1.9 Potential Uncertainties and Bias

The main sources of bias in the time series of abundance for the total spawning stock can be summarized as follows:

1) Incomplete and variable spatial coverage of the summer feeding distribution in all years which would lead to a negative bias;
2) Herding of mackerel into the trawl path by the trawl doors and the sweeps, in addition to the curved towing procedure, which could scare fish into the trawl path, and thus lead to positive bias;
3) Spatial and temporal variation in the portion of mackerel in the water column that is beneath the trawl (causes negative bias);
4) Escapement through the meshes or loss during hauling.

2.4.1.10 What are the Data Used for?

The data are used as a fishery independent time series of age-disaggregated abundance with precision, and are included as a tuning series in the stock assessment of NEA mackerel. To be used in the assessment, ICES requires time series to contain a minimum of five years of data. Such indices have previously been lacking for the mackerel assessment. The results provide valuable and highly needed information on abundance, distribution, and density of NEA mackerel. The survey has significantly improved the assessment and advice from ICES used for management of this highly valuable and ecologically important stock.

2.4.2 Bottom Trawl Surveys to Monitor Demersal Fish

2.4.2.1 Background

Bottom trawl surveys are typically used to estimate relative abundance of demersal (bottom) fishes. Many nations have bottom trawl surveys for demersal species that employ different survey designs or sampling design elements, for example, random survey design with fixed station locations (deep-water shrimp survey in the Skagerrak, run by IMR, Norway); stratified random survey with a 2-phase adaptive design and random station locations (Chatham Rise trawl survey, NIWA, New Zealand). This case study will focus on the North Sea International Bottom Trawl Survey (IBTS).

The North Sea IBTS surveys were begun in the 1960s with the primary objective to map the distribution of juvenile herring and investigate the links between herring nursery grounds and the adult populations (ICES, 1963). Initially, four nations took part in the survey, but the number of participating countries increased over the next decade. The objectives of the survey were expanded to include obtaining annual recruitment indices for the combined North Sea herring stocks and the survey area was also expanded to include the Skagerrak and Kattegat.

Although the survey objectives were focused on herring, data was collected for other species. Analysis of this other data showed that the survey objectives could be (and subsequently were) expanded to include providing indices for whitefish (gadoid) species. The survey went through several revisions from the early 1960s through to the current day, including combining multiple national surveys into one survey, survey area extension, reduction in seasonal coverage, and survey objective expansion (full history is in the IBTS survey manual; ICES, 2015a). Eight nations participate in the current North Sea IBTS surveys, which cover the entire North Sea (including the Skagerrak, Kattegat, eastern English Channel, and west of Shetland), take place during two seasons – winter (the first quarter of the year) and late summer/early autumn (quarter 3), and include all demersal and pelagic finfish, all chondrichthyans, and many invertebrate species.

2.4.2.2 Primary Objectives

The overarching purpose of the IBTS surveys is to provide ICES assessment and science groups with consistent and standardized data for examining spatial and temporal changes in (a) the distribution and relative abundance of fish and fish assemblages and (b) the biological parameters of commercial fish species for stock assessment purposes. This is achieved by the following objectives:

1) To determine the distribution and relative abundance of pre-recruits of the main commercial species with the view of deriving recruitment indices;
2) To monitor changes in the stocks of commercial fish species independently of commercial fisheries data;
3) To monitor the distribution and relative abundance of all fish species and selected invertebrates;
4) To collect data for the determination of biological parameters for selected species;

5) To collect hydrographical and environmental information;
6) To determine the abundance and distribution of late herring larvae (quarter 1 only).

2.4.2.3 Survey Design

A systematic grid survey design is used for the North Sea IBTS surveys, where the survey area is all of the North Sea, Kattegat, and west of Shetland to 200 m depth and the Skagerrak to 250 m depth (i.e., all of ICES Subarea 4 and Division 3a). The survey grid is based on ICES statistical rectangles, which are roughly 30 x 30 nautical miles (1 degree longitude x 0.5 degree latitude). These rectangles were already being used for fisheries management purposes, hence the convenience of continuing with their use when the survey was designed in the late 1960s.

Bottom Trawl Stations

Two stations are allocated to each rectangle for most of the survey area. Some rectangles with a high proportion of untrawlable ground are usually surveyed only once, others may have been allocated more than two stations for historical reasons pertaining to herring. The stations within a rectangle have typically been allocated to more than one nation (one per nation, two nations sampling per rectangle), although there are a few exceptions, for example, only Scotland samples west of Shetland, only Sweden surveys in the Skagerrak and Kattegat. A quasi-random sampling position is used; trawls are randomly chosen within each rectangle from known safe tow positions. The degree of randomness applied when determining sampling locations varies between nations. Trawl stations must be separated by 10 nmi and nations must avoid clustering stations between adjacent rectangles; this is to reduce positive serial autocorrelation and thereby maximize survey precision.

Sampling is only done during daylight hours because demersal fish move up in the water column (off the bottom) at night, reducing their availability to the survey. In addition, a standard protocol for fishing method, for example, tow time, maximum fishing depth, fishing speed, and trawl rigging were also set.

MIK Stations

During the quarter 1 survey, information is collected on the abundance, size, and distribution of herring larvae throughout the survey area. Four stations are allocated to each ICES rectangle, where stations must be at least 5 n mi inside the rectangle and separated by 10 n mi (if sampled in the same day) or 24 hours. To fit with the other survey components and because herring have fast escape behavior, MIKing is typically done at night. A strict protocol is followed for net material, rigging, deployment, and sorting of the collected sample (see manual, ICES, 2013a).

CTD Stations

A CTD station is coupled to all activities, where temperature and salinity are recorded for the entire water column. Traditionally, this has been taken using the CTD rosette, but nations are beginning to make the move to using CTDs that can be attached the net, which take water profiles from the entire trawl tow. This does not result in reduced data quality, can be argued to be truly the environment the fish are inhabiting at capture (and not just near the start or end of the tow), and saves time while out on survey.

2.4.2.4 Equipment
Bottom Trawl

A bottom trawl (see Chapter 3), the GOV (Grande Ouverture Verticale, GOV-trawl) with a rubber disc groundrope, was introduced as the standard gear. However, some countries continued to use gears other than the GOV in the third quarter IBTS until 1998. Because this survey has many nations taking part, the gear must remain standardized between all participants, therefore, the gear rigging and fishing method were standardized and documented in the survey manual (e.g., Figure 2.5; ICES, 2015a). Technological drift has occurred over time as nations have adjusted the gear to fit the conditions within their survey area. Procedure should have been to ensure all changes were well documented as they occurred, but this has unfortunately not taken place. The IBTS working group is now struggling with the question of how to continue with the 'standardized' survey.

In addition to the trawl, certain sensors must be used to monitor trawl geometry. These include net opening, wing spread, and door sensors, which measure opening from groundgear to headline, distance between the wings, and distance between the doors (as well as door pitch, roll, and depth).

MIK

The MIK net (or midwater ring net) is a reinforced version of the more delicate conical plankton sampling nets; this reinforcement is needed because of the harsh conditions in the North Sea in the first quarter. The ring frame is made of thick steel for strength and heaviness (although it is not uncommon for it to still be bent out of shape by the end of the survey) and the nets are strengthened with

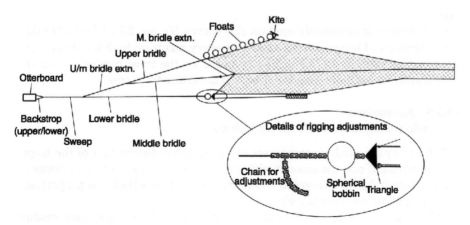

Figure 2.5 Standard rigging of the GOV trawl for use in the IBTS North Sea surveys. Modified from the IBTS survey manual (ICES, 2015a). *Source:* Artwork by R. Jakobsen.

Figure 2.6 MIK-M trawl for use in the North Sea IBTS first quarter surveys. The depth sensor (orange sensor) is visible on the metal ring. Small nets at side of MIK net are used to collect fish eggs. The flow meter is hanging in the center of the net (only the cord is visible). *Source:* RDM Nash, Institute of Marine Research, Norway.

nylon or canvas reinforcing straps (Figure 2.6). The net opening is 2 m diameter and the net is typically black (to help with the avoidance behavior of the larvae).

A flow meter is positioned in the center of the net so that volume of water sampled can be measured. Flow meters are calibrated to revolutions per meter before use. A depth (or preferably, a combined depth and temperature) sensor is also mounted on the ring, since sampling requires oblique tows to be taken, ending within 10 m of the bottom.

CTD
A CTD rosette is commonly used on this survey (Figure 2.7). The CTD takes measurements of temperature and salinity every meter on both the down- and upcast. Alternatively, a CTD can be mounted on the trawl headline to record temperature and salinity from the profile of the entire trawl tow.

2.4.2.5 Assumptions
The basic assumptions of this survey include:

- The survey area fully encompasses the geographic distribution of the target populations. If not, the amount within the survey area does not vary between years except due to changes in abundance (i.e., low availability is proportionally related to low abundance);
- Individuals are fully available to capture by the chosen gear (are neither attracted nor repelled);
- Individuals are fully available to capture during daylight hours (i.e., are on the bottom);

Figure 2.7 Example of a CTD rosette with 11 water collection canisters. *Source:* L. Nøttestad.

- Catchability of the different vessels is similar (or much less than the variability in abundance);
- Trawl performance remains constant under various conditions (e.g., weather, depth, bottom substrate);
- Size and species selection is constant under various conditions (e.g., weather, depth, bottom substrate, time of day, month/day of survey);
- Sites sampled are a representative selection from the survey area.

2.4.2.6 Computations

Because this is an international coordinated survey taking place in an ICES area, the ICES data center (DATRAS) holds the final (cleaned) data and carries out the basic computations (cpue, standard indices), which are then available for download by all users (general public, stock assessors, and industry). These computations are clearly outlined in ICES (2013b) and http://www.ices.dk/marine-data/Documents/DATRAS/Indices_Calculation_Steps_IBTS.pdf.

Indices are also estimated for several stocks using the method of Berg, Nielsen, and Kristensen (2014) and the R packages: *surveyIndex* (Berg, 2016) and *DATRAS* (Kristensen and Berg, 2010). The method for generating the indices is typically decided during the benchmark for that stock. Benchmarks are workshops designed to review, improve, and integrate with environment the assessment data and methodology under supervision of external (to ICES) peer reviewers. All stocks are required to undergo this process, at a maximum, every five years.

2.4.2.7 What are the Data Used for?

Once the survey is completed, the bottom trawl data are uploaded by the national (collecting) institutes to the ICES system (DATRAS), where they are freely available for use by anyone. While the data are cleaned prior to uploading, there are often small issues that any user needs to be aware of prior to using the data. All data has caveats, which need to be understood so that data are correctly analyzed. These caveats are clearly outlined on the DATRAS website, where the data are downloaded.

MIK data are not yet freely available to all; ICES is still working on a database system. Once this is complete, fish larvae data will be freely available.

All CTD data are uploaded to the international database at ICES and the World Ocean Database (ODB).

The data are used for the following (within the ICES framework):

- Stock assessments and forecasts (quota) for North Sea species;
- Estimating maturity ogives;
- Estimating stock weights-at-age;
- Generating recruitment indices for selected species;
- Determining distribution and abundance and any changes over time.

Outside of the ICES assessment and management needs, the data are used by many. A simple google search using the terms "DATRAS" and "North Sea" shows over 50 published papers in the last year.

2.5 Ecological Process Studies

2.5.1 Studying Diel Vertical Migration (DVM) of Mesopelagic Organisms Using Acoustics

Arved Staby, Institute of Marine Research, Norway

2.5.1.1 Background

A variety of marine organisms including zooplankton, planktivorous and piscivorous fish, and mesopelagic organisms perform diel vertical migrations (DVM). Two basic types of DVM patterns are identified: a type I migration describes the ascent of organisms to shallower waters just before the onset of night (dusk) and down to deeper daytime depths with day break (dawn), while the reverse applies to Type II migrations (Figure 2.8; Neilson and Perry, 1990). Light is thus considered an important and proximate factor in DVM, with changing light intensities governing the depth distribution of organisms and the timing and speed of their ascent to and descent from shallower water layers.

Different hypotheses explain the underlying value of vertical migrations in terms of (1) minimizing predation mortality by predator avoidance (Eggers, 1978), (2) distribution at temperatures that maximize growth (Wurtsbaugh and Neverman, 1988; Bevelhimer and Adams, 1993), (3) distribution at depths that correspond to high prey densities (Levy, 1990), and (4) regulating the depth according to the organism's internal hunger and satiation state (Pearre, 2003).

In the last decades hydroacoustics has increasingly become an important tool for aquatic ecologists to study the *in situ* behavior and vertical distribution of marine organisms (Bevelhimer and Adams, 1993; Kaartvedt *et al.*, 2009). During a diel period, mesopelagic organisms can occur in high densities within a narrow layer at different depths. In hydroacoustic terms, such densities are referred to as scattering layer (SL), and can be distinguished from other acoustic signals due to

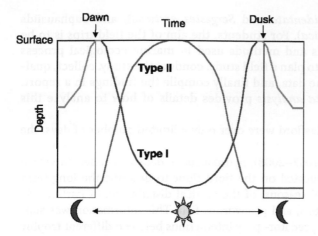

Figure 2.8 Example of Type I and II diel vertical migration. Type I diel vertical migration describes the movement of organisms from a bright surface environment to darker, deeper depths during daytime, while type II diel vertical migration describes the movement of organisms from a darker to a brighter environment during daytime. *Source:* A. Staby.

their much stronger echo (backscatter). Acoustic data is typically collected onboard moving or stationary vessels with hull mounted or submerged echo-sounders (Kaartvedt *et al.*, 2007) and, only fairly recently, have upward facing echosounders, positioned on the seafloor, been applied in behavioral studies of marine organisms (Axenrot *et al.*, 2004).

Acoustic observations alone are, however, uninformative if no prior information regarding the characteristics of the acoustic signals (targets) is available. Depth stratified pelagic trawl catches are therefore necessary to determine and verify the species composition of the acoustic signals at different times within a diel period. Specimens collected from the trawl catches can then be used to obtain data on length, weight, sex, gonad maturity stage, and stomach content.

Mesopelagic organisms are numerous in the open ocean (Irigoyen *et al.*, 2014), but are also numerous in coastal areas and deep West-Norwegian fjords (Salvanes and Kristoffersen, 2001; Salvanes, 2004). Masfjord, on the west coast of Norway, is approximately 20 km long, 1 km wide, and 490 m deep, and has been the location for numerous studies investigating the vertical distribution of SLs and the biology, ecology, and behavior of the mesopelagic organisms that inhabit it. The Department of Biology at the University of Bergen has conducted field courses for graduate students in marine biology in Masfjord over many years, allowing students to study the diel vertical migration behavior of fish and invertebrates. Given the relatively close proximity to land-based facilities, the fjord is also ideal for using an upward facing echosounder located on the bottom of the fjord to study the DVM behavior of the mesopelagic fish and large predatory fish.

2.5.1.2 Primary Objectives

The objectives of studying DVM in Masfjord are first and foremost to describe the dynamics of the vertical distribution of the mesopelagic community, which includes the fish species pearlside and glacier lanternfish, two shrimp species (*Pasiphaea multidentata* and *Sergestes arcticus*), and euphausiids (*Meganyctiphanes norvegica*). For students, the aim of the field trips is to be acquainted with the tools and methods used in marine ecological process studies, and to learn how to plan a field study, conduct the study, collect, quality control and analyze the data, and finally compile the findings in a report. Note that Chapter 5 *Data analysis* provides details of how to analyze this type of data.

Since most studies in Masfjord were over only a limited number of days, the knowledge regarding the behavior of DVM over long intermittent periods is limited. In a separate study (2007–2008), hydroacoustic data was collected with a stationary echo sounder located on the fjord floor to describe the long term dynamics, covering several seasons, of the vertical distribution of particularly pearlside and lanternfish SLs, and piscivorous fish. This information was supplemented with studies of predator–prey interactions between different trophic levels and dynamics of the vertical distribution behavior in relation to gradients in surface irradiation, water temperature, salinity, and oxygen.

2.5.1.3 Survey Design

The survey area is Masfjord. The design uses a (loosely) fixed sampling transect within the fjord that is revisited by the research vessel annually. The exact location is not revisited every year, but is rather a location in the general vicinity and at a similar depth.

The upward-facing echo sounder was located within the vicinity of the sampling transect at approximately 392 m depth. It was connected to a land-based laptop and power source via a 1.2 km long cable (Figure 2.9). All acoustic raw data was stored on the laptop where visualizations of the acoustic data (echograms) could be observed in real time.

2.5.1.4 Equipment

An overview of the various sampling gears and recording devices is shown in Figure 2.9.

The main sampling gears used in the Masfjord study were (1) a pelagic trawl rigged with a MultiSampler (Engås, Skeide, and West, 1997), which consists of three separate codends (20 mm mesh size) that can be opened and closed independently at different depths (described in Section 3.1.2.1) and (2) a MultiNet (Hydrobios) consisting of 5 nets (180 μm mesh size) that can be opened and closed independently (described in Section 3.1.2.4). With both the MultiNet and MultiSampler it is possible to collect uncontaminated depth stratified samples, meaning that the catch composition at a certain depth is not influenced by unwanted catch during hauling or heaving of the sampling gear.

A Seabird CTD recorded continuous temperature, salinity, and oxygen measurements between the surface and the bottom. Surface irradiation (photosynthetically active radiation, PAR at 400 to 700 nm) was measured with a calibrated LI-190 quantum sensor (LI-COR Biosciences) and stored on a LI-1400 data logger (LI-COR Biosciences).

During the field trips hydroacoustic data was collected with a hull mounted 38 kHz split beam echo sounder (SIMRAD). A calibrated upward-facing EK60

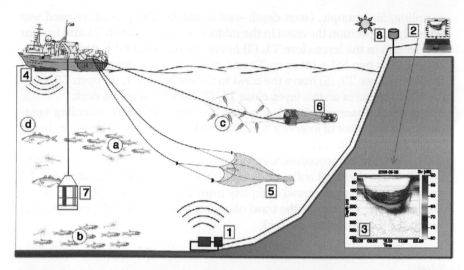

Figure 2.9 Schematic showing the various types of equipment used to sample the water column. (1) An upward facing EK60 38 kHz transducer located on the bottom of the fjord at 390 m depth sent sound waves toward the surface. This was connected to (2) a laptop on land, where (3) echograms could be followed in real-time. Data from (4) the hull-mounted echosounder was used to visualize acoustic targets/scattering layers at different depths. (5) A pelagic trawl combined with a MultiSampler was used to verify the species composition of the acoustic signals/targets (a, b, and d). (6) A plankton sampler collected depth-stratified plankton samples (c). (7) A CTDO recorded hydrographic data throughout the water column. (8) A light sensor measured radiation every 15 minutes. Large predatory fish (d) were also caught with rod and reel. *Source:* A. Staby.

38 kHz split beam echo sounder (SIMRAD) was used to collect acoustic recordings over a 15-month period in 2007–2008 (Figure 2.9).

2.5.1.5 Sampling
MultiSampler

For ease in describing the sampling method followed using the MultiSampler, the 3 trawl codends were encoded as: T1, the deepest trawl codend that opened first; T2, the middle codend; and T3, the shallowest codend that opened last.

The purpose of collecting trawl data in combination with hydroacoustic data is to describe (1) the species and length composition of acoustic targets, that is, scattering layers, and the depth variation of the scattering layers between day and night (referred to as "**follow scattering layers**") and (2) variation in abundance of mesopelagic organisms within depth intervals. The latter is done independently of acoustic observations and, if the aim is to monitor communities at fixed depth intervals, between depths over time and between years (referred to as "**fixed depth intervals**").

1) **Follow scattering layers**: This entails following the same layer of mesopelagic organisms over 24 hours. In Masfjord, there are generally only two scattering layers, referred to here as SSL1 (shallower layer) and SSL2 (upper layer). The start and end depth of towing was used to calculate the middle depth of

sampling, for example, (start depth–end depth)/2. The procedure used was typically: (1) position the trawl in the middle of SSL2, (2) open T1 and tow for 10 –15 min in the layer, close T1, (3) heave the trawl until it is positioned in-between the two SSLs, (4) open T2 and trawl for 10–15 minutes in-between the layers, close T2, (5) heave the trawl to the middle SSL1, (6) open T3, trawl for 10–15 minutes on this layer, close T3, (7) haul the trawl on deck.

2) **Fixed depth intervals**: Depth intervals are selected before sampling commences. The choice of intervals is determined by the research question. In the Masfjord, depth intervals were: 300–200 m (D1), 200–100 m (D2) and 100–0 m (D3). The trawling procedure was: (1) set the trawl at 300 m depth and open D1, (2) heave the trawl obliquely from 300 to 200 m depth and close D1, (3) open D2 and heave the trawl obliquely from 200 to 100 m depth, then close D2, (4) open D3 and heave the trawl obliquely from 100 m to the surface, after which the trawl is hauled on deck.

MultiNet

Each MultiNet net terminates in a numbered net-cup, which holds the sample. By using a rigid net-cup, fragile organisms are damaged less as the net is towed through the water. After sampling, but before the gear is taken onboard, the nets are hosed down with seawater so that plankton are flushed into the net-cups. The sieve on the side of each net-cup allows surplus water to drain away. The zooplankton samples from each of the net-cups are emptied into a bucket having the same number as the net-cup. Seawater is used to flush all organisms into the bucket. The MultiNet is then readied for the next tow and samples are taken to the lab for sorting.

Hook-and-line

Hook-and-line is used to sample larger fish. The fishing strategy depends on the research aims. For example, if changes in feeding activity at different depths throughout the day are to be studied, the entire depth range must be covered during each fishing period. A Data Storage Tag (DST), fixed to the terminal tackle, can be used to determine the exact capture depth. The DST will measure the depths (and temperature) at preset time intervals during fishing. By noting the capture time for each fish, the exact capture depth of each fish can be determined when the data have been downloaded from the DST. To keep track of fishing effort, each person that fishes has a unique identity code. Station information is recorded at the start of the fishing period.

2.5.1.6 Assumptions

All mesopelagic organisms in the path of the pelagic trawl are assumed to be captured. For following the scattering layer method, the catch is assumed to be representative of the sampled layers. For each of the fixed depth interval hauls, each trawl haul is assumed to capture a fixed proportion of the population present in the depth interval sampled, all organisms present can be caught, and all are equally likely to be captured. This implies that we assume that the catchability is constant over time and between hauls. Due to the lack of regular trawl sampling throughout the year, species composition of the SLs is assumed not to vary significantly between two sampling periods.

2.5.1.7 Computations

1) **Subsampling:** When equipment that samples a large volume or area is used, subsampling of the catch is common. Subsamples are taken when measuring all individuals would be too time consuming or when samples are very large (e.g., tons). The subsample is taken at random from a well-mixed total catch such that the proportion of each species will be the same in the subsample and the total catch. The proportions in weight and numbers in the subsample will then be the proportional to the total catch and calculations can be easily done. Procedures for subsampling are described in detail in Section 4.5.1.

2) **Acoustic data:** A detailed description of how the acoustic data was processed and analyzed is available in Kaartvedt *et al.* (2009). The raw **acoustic data** collected with the stationary echosounder were processed using the Sonar5-Pro (Balk and Lindem, 2009). Daily (24-hour) and monthly echograms were generated with MATLAB software and were assessed visually for variations in diel vertical migration patterns. A measure of the acoustic size of an organism is its target strength (*TS*), which is dependent on its length (Simmonds and MacLennan, 2005). By applying a *TS* threshold, acoustic targets from large fish can be differentiated from those of smaller fish. We applied automated target tracking (TT) to detect large individual fish using Sonar5-Pro, which assigns to the same target several consecutive echoes that satisfy preset track criteria. Tracks were counted in the upper 150 m, and assigned to 10 m vertical and 1 hour horizontal bins. To determine how SL depth varies with changing surface irradiation, acoustic data from the upper 250 m was divided into 10 minute (horizontal) and 1 m (vertical) bins. The mean volume backscattering strength (S_v) was calculated for each bin using the biomass calculation setup in Sonar5-PRO. The upper edge of SLs was defined as the depth where the average S_v dropped below −75dB.

3) **Surface irradiance:** A detailed description of how the surface irradiation and corresponding SL depth data was processed and analyzed is available in Staby and Aksnes (2011). The relationship between depth and surface radiation (E_0) is based on the assumption that fish at depth follow an isolume (E_{iso}), and that this depth is equivalent to Z_{iso} in the light attenuation equation (2.2):

$$E_{iso} = E_0 e^{-KZ_{iso}} \tag{2.2}$$

where E_{iso} is the downwelling irradiance at depth Z_{iso}, E_0 is surface irradiance, and K the light extinction coefficient. Solving equation (2.2) for Z_{iso}:

$$Z_{iso} = \frac{1}{K} \ln E_0 - \frac{1}{K} \ln E_{iso} \tag{2.3}$$

Equation (2.3) is another form of the linear function, where the slope a equals $1 \cdot K^{-1}$, x equals $\ln E_0$, and the y axis intercept b equals $1 \cdot K^{-1} \ln E_{iso}$ and is a constant. Provided that organisms that follow an isolume E_{iso} is constant, their depth Z_{iso} will be linearly related to changes in $\ln E_{iso}$.

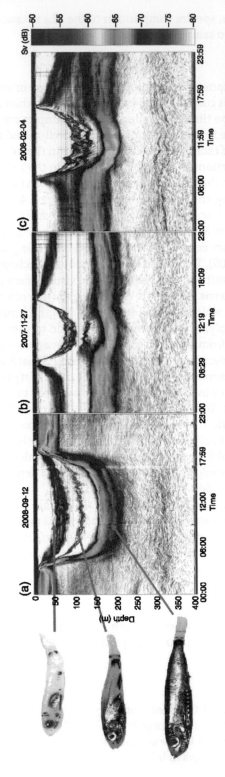

Figure 2.10 Movement of fish species in the water column over a 24-hour period; species in each layer were predominantly, from top to bottom, larvae, juvenile and adult pearlside and adult pearlside (*Maurolicus muelleri*). (a) Larvae, juveniles and adult pearlside are distributed at progressively deeper depths during daytime and aggregate in the upper water layer between dusk and dawn (type I migration). (b) Juvenile pearlside display nocturnal descents from the surface to approximately 50 m depth after dusk, while the deeper located adult fish (150–200 m) stay at depth throughout the diel period, with only a small proportion performing reverse diel migrations to 120 m. (c) Juvenile fish migrating to the surface stop their migration at different depths and descend again. *Source:* A. Staby.

2.5.1.8 Results

The major findings from the study with the stationary echosounder were that ontogenetic stages displayed seasonally varying migrating behaviors (Figure 2.10a, b, and c), including; (1) reverse diel migrations (type II DVM, Figure 2.10b), where, between November and January, a proportion of adult fish moved into slightly shallower and brighter depths; (2) arrested migrations (Figure 2.10c) characterized by ascending juvenile fish stopping their migration at different depths; and (3) seemingly light independent migrations at night to the surface. Some behaviors took place at the apparent lack of light, including early morning ascents and midnight sinking, or did not appear to be strictly governed by changes in light intensity, such as arrested migrations.

Catch data and counts of large predatory fish tracks revealed that mostly gadoid fish, dominated by saithe and pollack, performed type I DVM in summer, autumn, and winter months. In winter months, high counts of fish tracks at the surface corresponded with pearlside descending to deeper depths after dusk. This behavior suggests that predator evasion plays an important role in vertical distribution of pearlside. Stomach analyses of saithe and pollack suggested that these species fed on different-sized fish depending on their vertical distribution and time of day. Pearlside had ontogenetically varying feeding behavior. In autumn and winter months, adult fish did not perform DVM and had correspondingly empty stomachs. Juvenile fish did perform DVM and had full stomachs after returning from shallow waters at dawn.

As shown in Figure 2.11a the descent and ascent of pearlside SL corresponded largely with the start and end of civil twilight (when the sun is 6˚ below the horizon), respectively. SL depth and surface irradiance were negatively correlated, meaning that at high surface irradiation, pearlside SLs were positioned at deeper depths than at lower irradiation levels (Figure 2.11b). High coefficients of determination were obtained for the relationship between SL depth and surface irradiation, regardless of month, suggesting that the behavioral response of pearlside to changes in surface light intensity was consistent over time. However, monthly estimates of the preferred light level varied, suggesting that the light level fish choose to follow varies with age and state.

2.5.1.9 Potential Uncertainties and Bias

Automated TT is more objective than manual TT but it may include tracks that are not from individual fish, but rather from multiple targets or a random string of pings, introducing an error when using counts of fish tracks to determine their vertical distribution. Additionally, consecutive tracks may result from a single fish, thereby overestimating the number of tracks within an analyzed depth bin. The number of tracks detected will also be influenced by the distance (range) from the echosounder; the acoustic beam gets wider with increasing range and the likelihood of a fish swimming into a wider beam at the surface (long range) is much greater than close to the transducer (short range).

Some organisms may have a behavior that allows them to escape the trawls. Escape from the sampling gear depends on behavior, the swimming speed of the fish, the towing speed of the trawl, and the size of the individual. Fish may avoid the trawl by diving under, swimming over, or diving to the side of the gear when

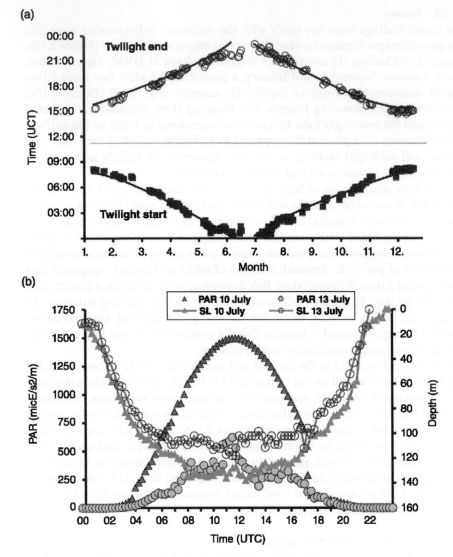

Figure 2.11 Migration of pearlside under different levels of light. (a) (upper panel) The arrival to and (lower panel) departure from the surface at civil twilight end and civil twilight start, respectively. Lines indicate the timing of the end and start of civil twilight, while symbols indicate when pearlside arrived or departed from the surface layer in a particular month. (b) Relationship between surface irradiation and SL depth over two days. The July 10 (blue triangles) was a clear day with high surface irradiation, while July 13 (circles) was overcast and had much lower irradiation recordings. Pearlside were distributed at deeper depths on July 10 compared to July 13. *Source:* A. Staby.

they sense the vibrations of the trawl in the water. Small fish may pass through the mesh of the net before they reach the codend. If the above is not taken into account, the species may be wrongly concluded to be absent. If only a few individuals are caught and catchability is assumed to be constant for size and species, abundance will be underestimated (as most will have escaped).

2.5.1.10   What are the Data Used for?

Mesopelagic fishes, in general, represent a large biomass in the mid-waters, forming prominent deep sound-scattering layers in world oceans (Gjøsæter and Kawaguchi, 1980), temperate deep fjords, and along continental slope areas (Salvanes and Kristoffersen, 2001, Salvanes, 2004). Their potential as new commercial resources was investigated back in the 1970s and estimated to one billion tonnes, which is equivalent to 10 times the global fish catch (Gjøsæter and Kawaguchi, 1980). Field studies in fjords since the mid-1980s demonstrated that vertically and diurnally migrating sound scattering layers (SSLs) represented movements of high biomasses of mesopelagic fish (Salvanes and Kristoffersen, 2001, Salvanes, 2004). The phenomenon of diurnal vertical migration has since then been used as a case study for a graduate field course in marine ecological field methods at the University of Bergen and onboard research vessels, with a focus on how to study this kind of phenomenon and the role of these organisms in the ecosystem. Irigoyien *et al.* (2014) estimated the global biomass estimate has increased by an order of magnitude since the 1970 Gjøsæter and Kawaguchi (1980) study, indicating that mesopelagic fish might be regarded as a potential protein resource for a growing human population.

2.5.2   Studying Barotrauma Impacts in Physoclistous Fish Species

Keno Ferter, Institute of Marine Research, Norway

2.5.2.1   Background

During fishing operations, a certain proportion of the catch is often discarded. In recreational fisheries, this practice is known as catch-and-release (C&R). C&R is practiced due to management regulations (i.e., regulatory C&R) and personal motivations (i.e., voluntary C&R) (Ferter *et al.*, 2013). The underlying principle of C&R is that the fish survive without significant negative long-term impacts after the C&R event (Arlinghaus *et al.*, 2007). However, C&R can lead to direct and indirect post-release mortality and sub-lethal impacts. Post-release mortality is species-specific and depends on several factors including, but not limited to, the anatomical hooking location, fighting time, water temperature, and capture depth (Bartholomew and Bohnsack, 2005). Physoclistous fish species (i.e., fish that have a closed swimbladder without a connection to the digestive tract), in particular, have increased post-release mortality and sub-lethal impacts when caught at deeper water depths (Alós, 2008; Hannah *et al.*, 2008a). The reason for this is that these fish develop barotrauma, which occurs when swimbladder gases expand due to ambient pressure reduction during forced ascent to the water surface (Hannah *et al.*, 2008b). Barotrauma can lead to a range of barotrauma signs, for example, exopthalmia ("pop-eyes"), everted stomachs, and gas bubbles in the blood, which can lead to post-release mortality (Rummer and Bennett, 2005). In addition, fish can become positively buoyant and thus have difficulties descending to capture depth, which makes them susceptible to avian predation. For some species, barotrauma signs have been shown to be reversible by using recompression devices that bring the fish down to capture depths (Butcher *et al.*, 2012). When management regulations require the release of fish suffering from barotrauma or voluntary C&R is practiced, it is important to evaluate the post-release survival and other impacts from barotrauma for these fish species. Atlantic cod (*Gadus morhua*) is one of these fish species and the impacts of barotrauma have been studied thoroughly in a complex field experiment (Ferter *et al.*, 2015), which is outlined below.

2.5.2.2 Primary Objectives

The main purpose of studying barotrauma impacts on Atlantic cod was to evaluate the effectiveness of current management regulations (i.e. minimum landing sizes) and common fishing practices in Norwegian marine recreational fisheries, and to develop best practice C&R guidelines. Atlantic cod is often caught at deeper water depths and it was therefore necessary to study the post-release survival of cod showing different barotrauma signs. This was achieved by studying the following primary objectives:

- Quantify the frequency of barotrauma signs with increasing capture depth;
- Quantify the percentage of floaters that cannot submerge to capture depth;
- Estimate post-release mortality of cod showing signs of barotrauma;
- Evaluate the use of recompression devices to reverse barotrauma impacts.

2.5.2.3 Survey Design

Atlantic cod were caught using rod and line at water depths down to 90 m. Only lure and hook types that minimize hooking injury were used because the study focused on capture depth-related effects on post-release mortality and not on C&R impacts due to hooking injury. Thus, only mouth-hooked cod without significant hooking injury were included in the experiment. All fish were caught on the bottom, and the exact capture depth for each cod was determined using a conventional echosounder.

Barotrauma Signs

Both external and internal barotrauma signs were investigated. External barotrauma signs (swollen coelomic cavity, gas release around the anal opening, gut eversion, stomach eversion, subcutaneous gas bubbles, and exophthalmia) were quantified immediately after capture for all fish that were used in the post-release mortality study (see section on post-release mortality). Internal barotrauma signs were quantified by dissection of additional fish that were killed immediately after capture. These fish were dissected to determine the presence or absence of gas in the venous blood system and to examine the swimbladder for its inflation status and the presence or absence of rupture holes.

Post-release Mortality

To quantify the proportion of floaters (i.e., positively buoyant cod that cannot submerge) and to study short-term post-release mortality, a containment study was conducted. Cod were caught on rod and line at depths between 7 and 90 m (treatment group) and in two-chamber cod pots (control group) from depths of

10 m. The control group was to account for mortality that might have been caused by experimental handling (e.g., tagging) and holding conditions. After capture and processing (i.e., investigation of barotrauma signs, length measurements, and individual tagging), the cod were released into a 5-m long cylindrical diving channel connected to a submersible cage (Figure 2.12). If the cod managed to dive into the submersible cage through the channel, they were classified as divers, if not, they were classified as floaters. After a maximum of 75 minutes

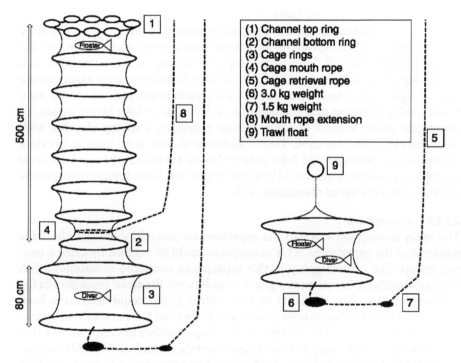

Figure 2.12 Diving channel and submersible cage setup. The channel was 500 cm in length and consisted of black knotless netting (20 mm mesh size). The channel was stabilized by eight rings which had a diameter ranging from 70 cm in the top (1) to 55 cm on the bottom (2). Directly above the lower end of the channel, a round net cage was attached, which consisted of black knotless netting. The cage was stabilized by two rings (90 cm in diameter (3)), which were 80 cm apart from each other, that is, the approximate cage height was 80 cm. At the top of the cage mouth was 75 cm of excess netting, which could be closed by a mouth rope (4) that was drawn through the net meshes. The cage was attached to the diving channel by putting the excess net over the lowest ring of the channel and tightening the cage mouth using the mouth rope (4) to a diameter of 30 cm. At the bottom of the cage, a cage retrieval rope (5) with a 3 kg weight (6) was attached (90 cm of rope between cage bottom and weight). Additionally, a 1.5 kg weight (7) was fixed 1 m from the 3 kg weight to buffer potential wave action after cage submersion. The cage could be retrieved to the surface by pulling on an extension (8) of the mouth rope that closed the cage mouth. After detachment of the cage from the channel, a small trawl float (9) was attached to the mouth rope to keep the cage upright after submersion. When the cage was lowered to the bottom, the retrieval rope was held at the surface using a large trawl float. This figure is reprinted with permission from Ferter *et al.* (2015a).

after capture, the submersible cage was retrieved to the surface, floaters were identified and transferred from the diving channel to the cage, and all cod (i.e., divers and floaters) were submerged to at least the capture depth. This procedure simulated the natural descent of divers and recompressed floaters to test if the use of recompression devices could be effective for cod to treat barotrauma. The cages were retrieved after a minimum of 72 hours and cod were classified as alive or dead.

2.5.2.4 Equipment

Rod and line and different **lures** or **baits** were required to catch fish for the experiment. An **echosounder** was used to determine the exact capture depth for each individual. This capture depth determination only works if the fish are caught at the bottom. If the fish are caught further up in the water column, a depth logger (e.g., data storage tag) has to be used to determine the exact capture depth. Captured fish were measured using a **length scale**, and individuals were marked using a **tagging gun** and **tags** (e.g., T-bar tags) for later identification. To investigate diving success and post-release mortality, a **diving channel with submersible cages** was used. **Two-chamber cod pots** were used to obtain a control group. Alternatively, **fyke nets** or **beach seines** can be used. External barotrauma signs were quantified by visual inspection, while internal barotrauma signs required the use of **dissection tools**.

2.5.2.5 Assumptions

The main assumption was that the experimental procedure did not affect the outcome of the experiment. This assumption could be verified by using a control group. The control group in this experiment consisted of uninjured fish without significant barotrauma signs that underwent the same procedure as the treatment group. Any mortality in the control group could thus have been ascribed to the experimental treatment and been used to interpret and/or correct mortality estimates in the treatment group. Another assumption was that the hooking did not cause any additional mortality which is why only mouth-hooked, non-bleeding cod were used, which had been shown to have very high survival in previous studies (Weltersbach and Strehlow, 2013; Ferter *et al.*, 2015b). Moreover, it was assumed that the 72 h holding period is long enough to cover the major part of mortality. A pilot study in this experiment and other studies (Humborstad, Davis, and Løkkeborg, 2009) showed that this time period is sufficient.

2.5.2.6 Computations

The frequency of barotrauma signs was analyzed using generalized linear models (see Ferter *et al.*, 2015a for details). The analysis of post-release mortality and the consideration of a control group are thoroughly described in Pollock and Pine (2007).

2.5.2.7 Results

The major finding was that the occurrence of external and internal barotrauma signs, in general, were significantly influenced by capture depth. All cod (floaters, divers, and control fish) survived the 72 h experimental observation period.

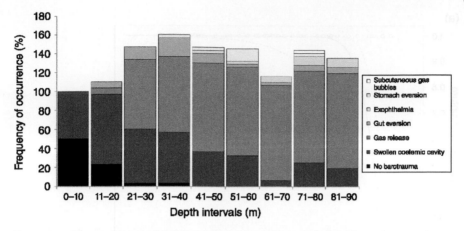

Figure 2.13 The occurrence of external barotrauma signs in angled cod from 0 to 90 m for each 10 m depth interval. As several individuals had more than one barotrauma sign, the sum of occurrence adds up to >100%. This figure is reprinted with permission from Ferter *et al.* (2015a).

A swollen coelomic cavity and gas release around the anus were the most frequently observed external barotrauma signs (Figure 2.13), while gas bubbles in the venous blood system and swimbladder rupture were common internal barotrauma signs (Figure 2.14). In total, 2.2% of the cod were classified as floaters and all floaters were caught in depths >50 m.

2.5.2.8 What are the Data Used for?

The results of this study have important implications for the practice and management of recreational Atlantic cod fishing in Norway. This study shows that short-term post-release mortality of Atlantic cod showing signs of barotrauma is low when the fish are otherwise not substantially injured and manage to submerge quickly after the release. Recreational fishers are thus encouraged to release cod independent of capture depth if the fish are under the minimum landing size, manage to submerge by themselves, and have no substantial hooking injury. Best practice guidelines should recommend the minimization of fighting time and air exposure so that the fish have enough energy to swim back to capture depth. Moreover, as all floaters that were recompressed in the submersible cages survived, recompression devices can be a useful tool to increase survival of floaters. Fisheries managers have been encouraged to discuss the implementation of suitable tools in the Norwegian marine recreational fishery. The sublethal effects (e.g., increased levels of stress hormone in the blood and behavioral changes) of barotrauma are not yet studied thoroughly. Thus, potential animal welfare issues need to be considered when cod from deeper capture depth are released which applies for voluntary C&R practice in particular.

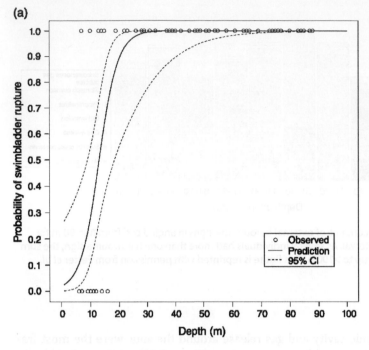

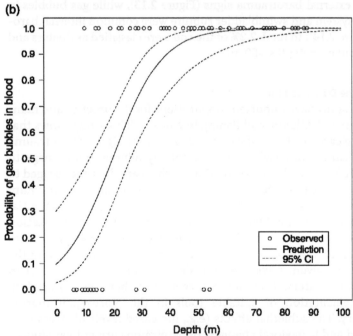

Figure 2.14 The probability of (a) swimbladder rupture and (b) gas bubble formation in the venous blood system (venous gas embolism) in angled cod with increasing capture depth. Points represent individual fish presence (1) and absence (0) data (many points overlap). The continuous lines show the model predictions and the dotted lines the range of the 95% confidence intervals. This figure is reprinted with permission from Ferter *et al.* (2015a).

References

Alós, J. (2008) Influence of anatomical hooking depth, capture depth, and venting on mortality of painted comber (*Serranus scriba*) released by recreational anglers. *ICES Journal of Marine Science*, 65(9), 1620–1625. DOI:10.1093/icesjms/fsn151

Arlinghaus, R., Cooke, S.J., Lyman J.*et al.* (2007) Understanding the complexity of catch-and-release in recreational fishing: an integrative synthesis of global knowledge from historical, ethical, social, and biological perspectives. *Reviews in Fisheries Science*, 15(1–2), 75–167. DOI:10.1080/10641260601149432

Axenrot, T., Didrikas, T., Danielsson, C. and Hansson, S. (2004) Diel patterns in pelagic fish behaviour and distribution observed from a stationary, bottom-mounted, and upward-facing transducer. *ICES Journal of Marine Science*, 61(7), 1100–1104. DOI:10.1016/j.icesjms.2004.07.006

Balk, H. and Lindem, T. (2008) *Sonar4 and Sonar5-Pro post processing systems.* Operator manual. Lindem Data Acquisition, Oslo.

Bartholomew, A. and Bohnsack, J.A. (2005) A review of catch-and-release angling mortality with implications for no-take reserves. *Reviews in Fish Biology and Fisheries*, 15(1–2), 129–154. DOI:10.1007/s11160-005-2175-1

Berg, C.W. (2016) surveyIndex: Calculate survey indices by age from DATRAS exchange data. R package version 1.0. https://github.com/casperwberg/surveyIndex (accessed June 21, 2017)

Berg, C.W., Nielsen, A. and Kristensen, K. (2014) Evaluation of alternative age-based methods for estimating relative abundance from survey data in relation to assessment models. *Fisheries Research*, 151, 91–99. DOI:10.1016/j.fishres.2013.10.005

Berge, J., Heggland, K., Lønne, O.J. *et al.* (2015) First records of Atlantic Mackerel (*Scomber scombrus*) from the Svalbard Archipelago, Norway, with possible explanations for the extension of its distribution. *Arctic*, 68(1), 54-61. DOI:10.14430/arctic4455

Bevelhimer, M.S. and Adams, S.M. (1993) A bioenergetics analysis of diel vertical migration by kokanee salmon, *Oncorhynchus nerka. Canadian Journal of Fisheries and Aquatic Sciences*, 50(11), 2336–2349. DOI:10.1139/f93-258

Bogstad, B., Pennington, M. and Vølstad, J.H. (1995) Cost-efficient survey designs for estimating food consumption by fish. *Fisheries Research*, 23, 37–46.

Buhl-Mortensen, L, Buhl-Mortensen, P, Dolan, M.F.J. and Holte, B. (2015) The MAREANO programme – A full coverage mapping of the Norwegian off-shore benthic environment and fauna. *Marine Biology Research*, 11(1), 4–17.

Butcher, P.A., Broadhurst, M.K., Hall, K.C. *et al.* (2012) Assessing barotrauma among angled snapper (*Pagrus auratus*) and the utility of release methods. *Fisheries Research*, 127, 49–55. DOI:10.1016/j.fishres.2012.04.013

Carlsson, D.M., Folmer, O., Kanneworff, P. *et al.* (2000) Improving the West Greenland trawl survey for *Pandalus borealis. Journal of Northwest Atlantic Fishery Science*, 27, 151–160.

Cochran, W.G. (1977) *Sampling Techniques*, 3rd edn, John Wiley & Sons, New York.

Connell, J.H. (1985) The consequences of variation in initial settlement vs. post-settlement mortality in rocky intertidal communities. *Journal of Experimental Marine Biology and Ecology*, 93(1–2), 11–45. DOI:10.1016/0022-0981(85)90146-7

Eggers, D.M. (1978) Limnetic feeding behavior of juvenile sockeye salmon in lake Washington and predator avoidance. *Limnology and Oceanography*, 23, 1114–1125.

Engås, A., Skeide, R., and West, C.W. (1997) The 'MultiSampler': a system for remotely opening and closing multiple codends on a sampling trawl. *Fisheries Research*, 29, 295–298

Ferter K., Hartmann K., Kleiven A.R., Moland E. and Olsen E.M. (2015b) Catch-and-release of Atlantic cod (*Gadus morhua*): post-release behaviour of acoustically pretagged fish in a natural marine environment. *Canadian Journal of Fisheries and Aquatic Sciences*, 72(2), 252-261. DOI:10.1139/cjfas-2014-0290

Ferter, K., Weltersbach, M.S., Humborstad, O.-B., *et al.* (2015) Dive to survive: effects of capture depth on barotrauma and post-release survival of Atlantic cod (*Gadus morhua*) in recreational fisheries. *ICES Journal of Marine Science*, 72(8), 2467–2481. DOI:10.1093/icesjms/fsv102

Ferter, K., Weltersbach, M.S., Strehlow, H.V. *et al.* (2013) Unexpectedly high catch-and-release rates in European marine recreational fisheries: implications for science and management. *ICES Journal of Marine Science*, 70(7), 1319–1329. DOI:10.1093/icesjms/fst104

Francis, R. (1984) An adaptive strategy for stratified random trawl surveys. *New Zealand Journal of Marine and Freshwater Research*, 18(1), 59–71. DOI:10.1080/00288330.1984.9516030

Francis, R.I.C. (2014) Replacing the multinomial in stock assessment models: a first step. *Fisheries Research*, 151, 70–84.

Gjøsæter, J. and Kawaguchi, K. (1980) *A review of the world resources of mesopelagic fish*. FAO Fisheries Technical Paper 193, 151 pp.

Godø, O. R., Pennington, M. and Vølstad, J. H. (1990) Effect of tow duration on length composition of trawl catches. *Fisheries Research*, 9, 165–179.

Gunderson, D.R. (1993) *Surveys of Fisheries Resources*, John Wiley & Sons, New York.

Hannah, R.W., Parker, S.J. and Matteson, K.M. (2008a) Escaping the surface: the effect of capture depth on submergence success of surface-released Pacific rockfish. *North American Journal of Fisheries Management*, 28(3), 694–700. DOI:10.1577/M06-291.1

Hannah, R.W., Rankin, P.S., Penny, A.N. and Parker, S.J. (2008b) Physical model of the development of external signs of barotrauma in Pacific rockfish. *Aquatic biology*, 3(3), 291–296. DOI:10.3354/ab00088

Humborstad, O.-B., Davis, M.W. and Løkkeborg, S. (2009) Reflex impairment as a measure of vitality and survival potential of Atlantic cod (*Gadus morhua*). *Fishery Bulletin*, 107(3), 395–402.

ICES (1963) International Young Herring Surveys. Report of working group meeting in IJmuiden, 26–27 March, 1963. ICES CM 1963/Herring Cttee: 101.

ICES (2013a) Manual for the Midwater Ring Net sampling during IBTS Q1. Series of ICES Survey Protocols SISP 2-MIK 2.18.

ICES (2013b) Report of the Workshop on Implementation in DATRAS of Confidence Limits Estimation. 10–12 May 2006, ICES Headquarters, Copenhagen. 39.

ICES (2013c) Report of the Workshop on Northeast Atlantic Mackerel Monitoring and Methodologies including Science and Industry Involvement (WKNAMMM), 25–28 February 2013, ICES Headquarters, Copenhagen and Hirtshals, Denmark. 33.

ICES (2014) Report of the Report of the Working Group on Widely Distributed Stocks (WGWIDE), 26 August–1 September 2014, ICES Headquarters, Copenhagen, Denmark. 938.

ICES (2015a) Manual for the International Bottom Trawl Surveys. Series of ICES Survey Protocols SISP 10 - IBTS IX. 86.

ICES (2015b) Report of the Working Group of International Pelagic Surveys (WGIPS), 19–23 January 2015, ICES Headquarters, Copenhagen, Denmark. 279.

ICES (2016) Report of the Working Group on Widely Distributed Stocks (WGWIDE), 31 August–6 September 2016, ICES Headquarters, Copenhagen, Denmark. 500.

Irigoyien X., Klevjer, T.A., Røstad, A. *et al.* (2014) Large mesopelagic fishes biomass and trophic efficiency in the open ocean. *Nature communications*, 5. DOI:10.1038/ncomms4271

Jolly, G. and Hampton, I. (1990) A stratified random transect design for acoustic surveys of fish stocks. *Canadian Journal of Fisheries and Aquatic Sciences*, 47(7), 1282–1291. DOI:10.1139/f90-147

Kaartvedt, S., Klevjer, T.A., Torgersen, T. *et al.* (2007) Diel vertical migration of individual jellyfish (*Periphylla periphylla*). *Limnology and Oceanography*, 52(3), 975–983. DOI:10.4319/lo.2007.52.3.0975

Kaartvedt, S., Røstad, A., Klevjer, T.A. and Staby, A. (2009) Use of bottom-mounted echo sounders in exploring behavior of mesopelagic fishes. *Marine Ecology Progress Series*, 395, 109–118. DOI:10.3354/meps08174

Kish, L. (1965) *Survey Sampling*. John Wiley & Sons, New York, IX + 643.

Kish, L. (1995) Methods for design effects. *Journal of Official Statistics*, 11(1), 55–77.

Kristensen, K. and Berg, C. (2010) DATRAS: R package version 1.0. Read and convert raw data obtained from http://datras.ices.dk/Data_products/Download/Download_Data_public.aspx. (accessed June 21, 2017)

Levy, D.A. (1990) Sensory mechanism and selective advantage for diel vertical migration in juvenile sockeye salmon, *Oncorhynchus nerka*. *Canadian Journal of Fisheries and Aquatic Sciences*, 47(9), 1796–1802. DOI:10.1139/f90-204

Morehead, S., Montagna, P. and Kennicutt, M.C. (2008) Comparing fixed-point and probabilistic sampling designs for monitoring the marine ecosystem near McMurdo Station, Ross Sea, Antarctica. *Antarctic Science*, 20(5), 471–484. DOI:10.1017/S0954102008001326

Neilson, J. and Perry, R. (1990) Diel vertical migrations of marine fishes: an obligate or facultative process? *Advances in marine biology*, 26, 115–168. DOI:10.1016/S0065-2881(08)60200-X

Nelson, G.A. (2014) Cluster sampling: a pervasive, yet little recognized survey design in fisheries research. *Transactions of the American Fisheries Society*, 143, 926–938.

Nøttestad, L., Diaz, J., Penã, H. *et al.* (2016a) Feeding strategy of mackerel in the Norwegian Sea relative to currents, temperature, and prey. *ICES Journal of Marine Science*, 73(4), 1127–1137. DOI:10.1093/icesjms/fsv239

Nøttestad, L., Utne, K.R., Óskarsson, G.J. *et al.* (2016b) Quantifying changes in abundance, biomass, and spatial distribution of Northeast Atlantic mackerel (*Scomber scombrus*) in the Nordic seas from 2007 to 2014. *ICES Journal of Marine Science*, 73(2), 359–373. DOI:10.1093/icesjms/fsv218

Paine, R.J. (1974) Intertidal community structure: Experimental studies on the relationship between a dominant competitor and its principal predator. *Oecologia*, 15(2), 93–120. DOI:10.1007/BF00345739

Pearre, S.J. (2003) Eat and run? The hunger/satiation hypothesis in vertical migration: history, evidence and consequences. *Biological Reviews*, 78(1), 1–79. DOI:10.1017/S146479310200595X

Pennington, M. and Vølstad, J.H. (1991) Optimum size of sampling unit for estimating the density of marine populations. *Biometrics*, 47, 717–723.

Pennington, M. and Vølstad, J.H. (1994) Assessing the effect of intra-haul correlation and variable density on estimates of population characteristics from marine surveys. *Biometrics*, 50, 725–732.

Pennington, M., Burmeister, L.-M. and Hjellvik, V. (2002) Assessing the precision of frequency distributions estimated from trawl-survey samples. *Fishery Bulletin*, 100, 74–81.

Pollock, K. and Pine, W. (2007) The design and analysis of field studies to estimate catch-and-release mortality. *Fisheries management and ecology*, 14, 123–130.

Rummer, J.L. and Bennett, W.A. (2005) Physiological effects of swim bladder overexpansion and catastrophic decompression on red snapper. *Transactions of the American Fisheries Society*, 134(6), 1457–1470. DOI:10.1577/T04-235.1

Salehi, M., Moradi, M., Brown, J.A. and Smith, D. (2010) Efficient estimators for adaptive stratified sequential sampling. *Journal of Statistical Computation and Simulation*, 80(10), 1163–1179. DOI:10.1080/00949650903005664

Salvanes, A.G.V. (2004) Mesopelagic fish, in *The Norwegian Sea Ecosystem* (ed. H.R. Skjoldal) Tapir Academic Press, Trondheim, pp. 301–314.

Salvanes, A.G.V. and Kristoffersen, J.B. (2001) Mesopelagic fish (life histories, behaviour, adaptation), in *Encyclopedia of Ocean Sciences* (eds: J. Steele, S. Thorpe and K. Turekian), Academic Press Ltd., San Diego, pp. 1711–1717.

Schonbeck, M.W. and Norton, T.A. (1980) Factors controlling the lower limits of fucoid algae on the shore. *Journal of Experimental Marine Biology and Ecology*, 43(2), 131–150. DOI:10.1016/0022-0981(80)90021-0

Simmonds, J. and MacLennan, D.N. (2005) *Fisheries Acoustics: Theory and Practice*, 2nd edn. Blackwell Science, Oxford.

Skinner, C.J., Holt, D. and Smith, T.M.F. (1989) *Analysis of Complex Surveys.* Wiley & Sons, Chichester.

Smith, E.P. (2002) BACI design. *Encyclopedia of environmetrics*. Wiley & Sons, Chichester.

Staby, A. and Aksnes, D.L. (2011) Follow the light – diurnal and seasonal variations in vertical distribution of the mesopelagic fish *Maurolicus muelleri*, *Marine Ecology Progress Series*, 422, 265–273.

Tirasin, E.M. and Jørgensen, T. (1999) An evaluation of the precision of diet description. *Marine Ecology Progress Series*, 182, 243–252.

Trenkel. V.M., Huse, G., MacKenzie, B. *et al.* (2014) Comparative ecology of widely distributed pelagic fish species in the North Atlantic: implications for modelling climate and fisheries impacts. *Progress in Oceanography*, 129, 219–243. DOI:10.1016/j.pocean.2014.04.030

Warren, W.G. (1994) The potential of sampling with partial replacement for fisheries surveys. *ICES Journal of Marine Science*, 51, 315–324.

Weltersbach, M.S. and Strehlow, H.V. (2013) Dead or alive – estimating post-release mortality of Atlantic cod in the recreational fishery. *ICES Journal of Marine Science*, 70(4), 864–872. DOI:10.1093/icesjms/fst038

Wurtsbaugh, W.A. and Neverman, D. (1988) Post-feeding thermotaxis and daily vertical migration in a larval fish. *Nature*, 333, 846–848. DOI:10.1038/333846a0

Further Reading

Baliño, B. and Aksnes, D.L. (1993) Winter distribution and migration of the sound scattering layers, zooplankton and micronekton in Masfjorden, western Norway. *Marine Ecology Progress Series*, 102, 35–50.

Staby, A., Røstad, A. and Kaartvedt, S. (2011) Long-term acoustical observations of the mesopelagic fish *Maurolicus muelleri* reveal novel and varied vertical migration patterns. *Marine Ecology Progress Series*, 441, 241–255. DOI:10.3354/meps09363

3

Sampling Gears and Equipment

Anne Gro Vea Salvanes*, Henrik Glenner*, Dag L. Aksnes, Lars Asplin,
Martin Dahl, Jennifer Devine, Arill Engås, Svein Rune Erga,
Tone Falkenhaug, Keno Ferter, Jon Thomassen Hestetun,
Knut Helge Jensen, Egil Ona, Shale Rosen and Kjersti Sjøtun

3.1 Sampling Organisms

3.1.1 Direct Observations

3.1.1.1 Littoral Zone Methods

Littoral transects: Since the littoral zone is horizontally stratified, vertical transects through the upper and down into the lower littoral zone are a relatively quick way of describing the zonation at a particular point along the shore by mapping the different zones of dominating organisms at study localities. Commonly, a rope or measuring tape is placed perpendicular from a fixed point in the upper part of the zone, down into the lower zone, and the positions of the dominating species along the transect are registered. The method is often supplemented with GPS equipment to accurately measure the vertical heights of the dominating species. If one has access to a hydrographical database with accurate measurements or calculations of tidal levels above Chart Datum and, if the time of each measurement is noted, all measurements can be related to Chart Datum (Figure 3.1). In this way, the levels and extensions of the different zones can be compared between years. If desired, transects can be extended into the subtidal part of the littoral zone through scuba diving or by use of a ROV or campod.

Sample squares/quadrat-based methods: For a littoral biodiversity survey where a high precision level is needed, detailed community analyses are necessary. This can be achieved using sampling plots or quadrats and by making a detailed recording of organisms present inside each sample plot.

An example of sample squares made of steel with a standard size of $0.25\,\mathrm{m}^2$ is shown in Figure 3.2. The second square has a grid of 25 $0.01\,\mathrm{m}^2$ subsquares, which is then placed on top of the first square to make it easier to record sessile species.

---

* Lead authors; co-authors in alphabetical order.

Marine Ecological Field Methods: A Guide for Marine Biologists and Fisheries Scientists,
First Edition. Edited by Anne Gro Vea Salvanes, Jennifer Devine, Knut Helge Jensen,
Jon Thomassen Hestetun, Kjersti Sjøtun and Henrik Glenner.
© 2018 John Wiley & Sons Ltd. Published 2018 by John Wiley & Sons Ltd.

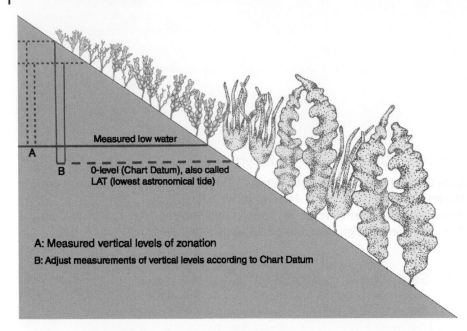

Figure 3.1 A schematic figure showing macroalgal zones in an intertidal, demonstrating the principles for measuring vertical levels and zones in relation to Chart Datum (0-level or LAT). *Source:* Artwork by R. Jakobsen.

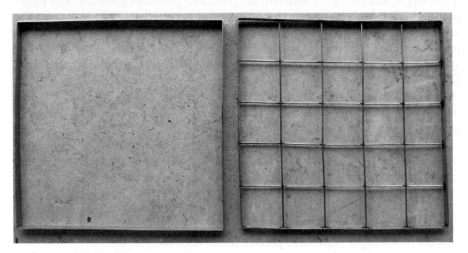

Figure 3.2 An example of a steel quadrat and grid overlay used for sampling. *Source:* J.T. Hestetun.

Abundance data of species within the sample square can be collected in different ways. A common way of recording species in the intertidal is to estimate the percentage cover of sessile (attached) organisms and to record the number of vagile (moving) animals within defined sample squares (Figure 3.3). The sample squares can be placed in fixed positions using permanent bolts in the rock at the

Figure 3.3 Use of sample quadrats in the field. The sample frames are positioned by attaching the underlying frame to two permanently fixed bolts in the rock. *Source:* J.T. Hestetun.

locality, which is normally done in monitoring surveys extending over a longer time period. After placing the sample square into position, counts of the number of organisms for easily countable species (such as gastropods, crustaceans) are made inside the square. A lower size limit for recording individuals, for example, 3 mm, is typically decided before beginning. If an organism is inside the sample square initially, but later moves out, it should be recorded. Slow-moving individuals like limpets should be recorded as present if more than half of their body is inside the sample square. Macroalgae and sessile animals are recorded as percent coverage of the substratum. Since there are 25 small squares in the grid, each square represents 4% total cover. Examples of sessile animals are barnacles, bryozoans, hydroids, and some polychaetes like *Spirorbis spirorbis*. Large animals, which have a limited ability to move, like *Mytilus edulis*, can be either counted or recorded as percent surface coverage; once a decision is made, one must stick to the chosen procedure. Information about how each taxa is recorded should be written down on the recording form.

Epiphytes covering the surface of macroalgae (e.g., other macroalgae and sessile animals) should be recorded separately. In many cases, one needs to remove the overlying vegetation, consisting of large fucoids or kelps and complete a second estimation of the underlying organisms. In this case, percentage cover of all species will often exceed 100%. Coverage of smaller, single, stationary animals or algae is typically recorded as presence/absence, encoded as 1% in the dataset. Individuals that are difficult to identify to species level *in situ* should be put in vials marked with site and square number and brought to the laboratory for later identification. In some cases, it is difficult to identify an organism to species level and one must settle for genus or even a higher taxon level.

For species covering the substratum in small patches, estimating percent coverage can be particularly difficult. To make the result as accurate as possible, two people can make separate estimates, which are then averaged. Another method is to use stereo photographing and analyze the photos later. This is more time consuming, but will give a more precise result.

There are also other ways of recording abundance of species in the sample squares (Meese and Tomich, 1992). One method is to harvest all biomass within the sample frame, sort into species, and measure the weight of the different species. This should not be done if the survey is part of a time series because this is a destructive method. Another alternative is to use the frequency recording method. Here, all species at specific predetermined points within the frame are recorded. Predetermined points are produced by crossing points of the grid overlying the square frame. This method requires a relatively high number of points in order to avoid over- or under-estimate rare species.

3.1.1.2 ROV Sampling

ROVs (remotely operated vehicles) have become a very common tool in many marine biological surveys (Figure 3.4). The ability to capture high definition video, coupled with the possibility of using different types of sampling equipment, give ROVs a flexibility unmatched by most types of sampling equipment.

Live video feed allows the ROV to be positioned and to observe points of interest directly. The video data can be used for many purposes (see Section 3.1.1.3). A ROV can be fit with a variety of implements and sampling gear. Mechanical arms with manipulators are common, as is one or several storage containers.

Figure 3.4 A remotely operated vehicle (ROV). *Source:* J.T. Hestetun.

Another mainstay of biological sampling is the suction sampler, which draws water through a tube into a container, with more advanced models having several revolving chambers. ROVs allow the simultaneous precision sampling of other parameters such as temperature, depth, and *in situ* imaging, making them valuable in describing the conditions of organisms and communities.

3.1.1.3 Video/image-based Methods

To quantify samples from hard-bottom habitats constitutes a particular challenge because the substrate, unlike soft-bottom habitats, cannot be retrieved. However, by using a campod or an ROV with video recording capabilities, quantitative, or at least, semi-quantitative studies can be conducted. Stored video footage can be analyzed in detail either manually by trained personnel or using specialized software that pieces together mosaics of the hard-bottom sites (e.g. 1x1 m) and, to a certain degree, identify and quantify characteristic species. This method has been employed in studies with a limited focus, but the resulting list of species will contain only a restricted number of predetermined species (the ones that can be identified from the video image alone; see also Section 3.1.4.2 AUVs).

3.1.1.4 Manned Submersibles

The use of manned submersibles has long been a high-profile method to directly observe deep-sea habitats. Increasingly, usage of manned submersibles is being complemented and partially supplanted by ROVs and AUVs, which allow the same benefits with less use of resources.

3.1.1.5 Scuba Diving

For certain types of studies in the sublittoral and shallow waters, scuba diving allows targeted collection of samples of interest. Scuba diving may be employed together with quadrat analysis or other methods.

3.1.2 Active Gears

Active gears capture species by actively chasing the species or actively hunting the target species.

3.1.2.1 Sampling Trawls (Midwater and Bottom)
Arill Engås and Shale Rosen, Institute of Marine Research, Norway

Trawls are cone-shaped nets used to sample mobile species over relatively large areas. They are used worldwide to sample throughout the water column, from seabed to surface. Trawl samples can be used to collect samples for determining biological parameters such as age, sex, genetics, and diet, and to estimate distribution and abundance of different species and sizes. To capture an organism, the trawl must be towed at a speed greater than the target species can swim.

Pelagic or midwater trawls are typically used to capture pelagic or benthopelagic species. A pelagic trawl consists of pelagic trawl doors (also called otter boards), bridles, and a cone-shaped trawl net, where the large opening (front of the trawl) is framed by a headrope (top) and a footrope (bottom). The trawl doors (otter boards) exert outward force as they are pulled through the water, opening

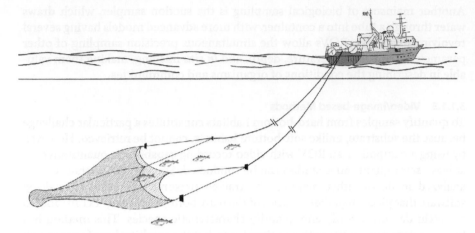

Figure 3.5 Illustration of a pelagic trawl and its components. For some pelagic trawls, lift is created along the headrope by a kite or additional cable connected to the vessel. *Source:* Artwork by R. Jakobsen.

the trawl horizontally. Vertical opening is created by the opposing forces of weights on the lower bridles and/or footrope and lift on the headrope by the addition of floats, kites, or tension from a net sonar cable connected to the vessel (Figure 3.5).

The trawl position in the water column can be adjusted by varying the length of trawl warp (wire connecting trawl to the vessel). Pelagic trawls often have openings that measure tens of meters in height and width. The large opening is necessary because fish can escape the approaching trawl by diving below, swimming up, or swimming to the sides.

Pelagic trawls are often constructed with large meshes in the forward section, which can be tens of meters in size, and act to herd fish rather than capture by sieving. The aft portion of the trawl, called the codend, is constructed of mesh sizes that are sufficiently small that target sizes cannot escape. Pelagic trawls are typically towed at higher speeds than other sampling gears; speeds of 2.5 m sec^{-1} are not uncommon when targeting fast swimming species such as Atlantic mackerel (*Scomber scombrus*).

Demersal or bottom trawls are commonly used for sampling the bottom living organisms (primarily fish). A demersal trawl is similar in design to a pelagic trawl, but the trawl doors, lower bridles, and groundline are designed to be in constant contact with the seabed (Figure 3.6). Groundgear is often attached to the footrope to prevent it from snagging or being torn on the seabed. In areas with relatively smooth seabed, the groundgear can be as simple as thick rope. When operating in rocky habitats, rubber disks of up to 50 cm diameter may be placed along the groundgear. This is commonly referred to as "rockhopper gear."

Demersal trawls made for research surveys tend to be smaller than pelagic trawls, with a typical net opening of 2–4 m height. The net opening is maintained by floats attached to the headline and a weighted groundrope. Bottom trawls used for commercial fishing have much larger net openings, some of which can engulf a 747 airplane with room to spare. The ground gear is in contact with

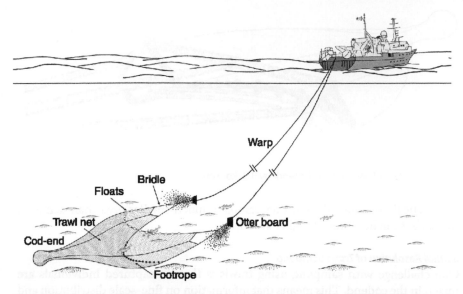

Figure 3.6 Illustration of a bottom trawl and its components. Mud clouds created by the otter boards (trawl doors), lower bridles, and footrope serve to herd fish into the opening of the trawl. *Source:* Artwork by R. Jakobsen.

the seabed, which provides a barrier to escape below the trawl. However, some species are able to escape below the groundrope, usually between gaps in the bobbins, which vary in size from small rubber discs to large metal balls up to 0.5 m in diameter. Demersal trawl doors (otter boards) serve a dual purpose; in addition to spreading the trawl horizontally, they create both noise and sand clouds as they travel across the seabed, which in combination with the bridles, herds fish into the path of the trawl opening. The effective fishing area can, therefore, be as wide as the distance between the trawl doors. In contrast to pelagic trawls, mesh size for the trawl netting in the front portion of demersal trawls rarely exceeds 200 mm. As with pelagic trawls, codend mesh size is chosen to retain the target size individuals. Typical trawling speed for demersal trawling is in the range of $1–2\,\mathrm{m\,sec^{-1}}$.

Beam trawls are typically used to sample flatfish and shrimp species on the seabed. The headline is attached to a rigid steel frame comprised of a horizontal beam running across the top and beam heads at either side. Steel "shoes" attached to the underside of the beam head slide along the seabed. The ends of the ground-line are connected to the shoes. Depending on the type of seabed and target species, tickler chains or chain mats are often mounted between the shoes ahead of the groundline (Figure 3.7). The rigid frame provides both horizontal and vertical spread to the netting, so trawl doors are not needed and only a single trawl warp is necessary. This makes beam trawls particularly suited to sampling from small vessels. Mesh size of the netting and codend varies with the target species and sizes.

Handling considerations for the rigid steel beam mean that beam trawls are typically smaller than demersal trawls. With horizontal openings of less than 10 m

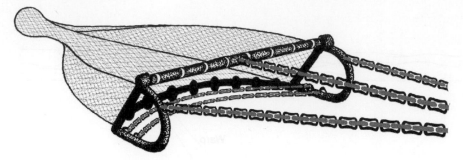

Figure 3.7 Beam Trawl. *Source:* Artwork by R. Jakobsen.

and vertical openings of <1 m they sample a much smaller area than demersal trawls, but can be towed at higher speed.

Spatial Resolution of Trawl Sampling

One challenge with sampling using trawls is that all captured individuals are mixed in the codend. This means that information on fine-scale distribution and species overlap is lost as it is impossible to know where along the trawl path an individual was captured. This is a problem particular to the pelagic zone, where the trawl may have passed through a depth range of several hundred meters. To determine when individuals were captured during trawling, in-trawl cameras and MultiSamplers can be used.

The Cam-trawl system (Williams, Towler, and Wilson, 2010) and Deep Vision (Figure 3.8; Rosen *et al.*, 2013; Rosen and Holst, 2013) are two examples of **in-trawl stereo camera systems** that can verify species and size as organisms pass into the codend of a trawl. The use of stereo imagery makes it possible to calculate size of the passing objects. These technologies also make it possible to conduct surveys with an open codend, which is useful in situations where catches would be too large to handle on a research vessel or result in unacceptable levels of mortality (for example when surveying rare or threatened species). An example of Deep Vision results showing species and size distribution throughout an oblique pelagic trawl haul is presented in Figure 3.9.

(a)   (b)

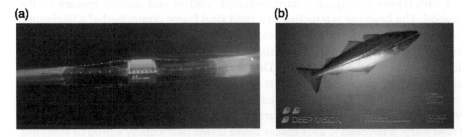

Figure 3.8 (a) DeepVision system placed between the trawl and codend of a pelagic trawl. (b) Example of a 100 cm Atlantic cod (*Gadus morhua*) taken by the DeepVision system. Precise time and position are indicated in the white overlay text. *Source:* With permission from Scantrol DeepVision.

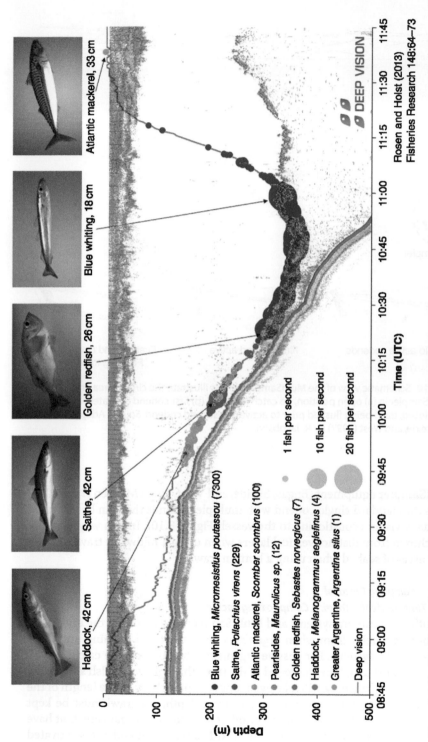

Figure 3.9 Vertical position, length, and density of fish in the water column as measured by DeepVision system, overlaid on an acoustic echogram. Trawl path is indicated by the thin gray line, while location of fish is by colored circles with diameter scaled to indicate the number of fish passing per second. Seabed is indicated by the sloping red band. Images above the figure are the specific fish indicated on the depth profile. *Source:* With permission from Rosen and Holst (2013).

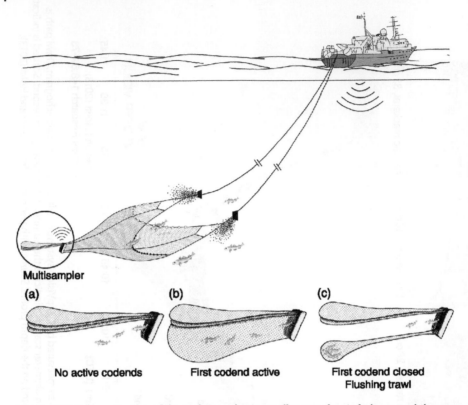

Multisampler

(a)   **(b)**   **(c)**

No active codends   First codend active   First codend closed
                                                    Flushing trawl

Figure 3.10 Schematic view of the MultiSampler. Insets illustrate the circled area and show the MultiSampler in (a) open position, no catch retained; (b) first codend activated; and (c) first codend closed, trawl being flushed prior to activating second codend. *Source:* Adapted from Engås, Skeide, and West (1997) by R. Jakobsen.

MultiSampler equipment (Engås, Skeide, and West, 1997; Madsen *et al.*, 2012) replaces the standard single codend with multiple codends that can be activated sequentially via an acoustic link to the vessel (Figure 3.10). In this way, samples can be chosen from distinct vertical layers when using a midwater trawl or from specific areas of seabed when using a bottom trawl.

3.1.2.1.1 Sampling Protocol
Passage Time Inside the Trawl: A Complicating Factor
Since both trawl camera systems and MultiSamplers are placed at the end of the trawl, there is a delay from when organisms enter the trawl opening until they pass through the systems. For passive organisms. the delay can be calculated by dividing the trawl length by the speed of the trawl through water. Fish swimming in the direction of trawling will take even longer to pass through the length of the trawl. When attempting to sample at a distinct depth, the trawl must be kept steady, with all codends closed for a period of time to flush organisms that have entered at the previous depth (before the next MultiSampler codend is activated or images are analyzed).

Methods of Fishing (Midwater)

The objectives of the study determine how the trawl is fished. One method is to fish at **fixed depth intervals**. This is commonly used if the aim is to determine species distributions or define community composition. If samples are taken throughout the day and night, they can be used to study diurnal changes in abundance or composition. Another is to fish on the echo-layers that are detected using acoustics. This is common if the intention is to identify what is in aggregations recorded by acoustics. This is commonly referred to as target registration towing.

Target Registration Towing

This entails following the same aggregations of fish, whether they are distributed in layers, patches, or schools. This allows investigating whether the different layers/patches/schools differ with respect to species or size composition and whether any observed differences persist with, for example, time of day. Sampling can also be conducted between layers to determine species occurrence as larger predators are often found between layers.

3.1.2.2 Beach Seine

Beach seine fishing is a method of fishing that utilizes a **seine** or **drag net** to sample shallow, nearshore areas where the sublittoral slope is not too steep. A beach seine is operated from the shore, often with the help of a small boat. The gear is composed of a net (or bag) and long wings that terminate in long ropes for towing the seine to the beach. The headrope with floats is on the surface and the footrope with weights is in permanent contact with the seabed. The seine acts as a barrier that prevents fish from escaping the area enclosed by the net. Seine nets, deployed from the shore or from a boat, are set in a ring and hauled towards the beach with the bottom line dragging over the bottom. There are two types of seine nets: those without a bag and those with a bag (Figure 3.11).

Figure 3.11 Deployment of a beach seine. *Source:* A.G.V. Salvanes.

The beach seine without a bag has a central part with smaller meshes. No advanced gear handling equipment is required for fishing operations, but a number of people (at least four) are needed to haul the seine to the shore. Since hauling is over the bottom, the beach seine should preferably be used on seabeds with relatively flat (smooth) bottom and without large boulders to avoid damage. A wading suit is recommended for remaining dry while removing boulders or other obstructions should the seine become stuck when hauling.

3.1.2.2.1 Sampling Protocol

One towing line is either fastened on shore or the line is held by at least two people. Then the rest of the gear (line, first wing, bag, second wing, and second line) is set in a wide arch, either with the help of a boat or by swimming/wading, and brought back to the shore where two other people are standing. The drag lines are then hauled simultaneously by the people on the beach. Fish are herded in front of the bag or net. The ground rope must reach the beach first so that the gear captures the fish (i.e., the fish do not escape underneath the net).

Target species are small sublittoral crustaceans and fish. Shallow waters close to the shore are often spawning and nursery grounds. Beach seining can disturb the breeding activities (if at spawning time) or catch juveniles of commercial species. For these reasons, the beach seine is regulated or restricted by law in a number of countries. Sampling using beach seine may therefore require permits from the fishing authorities, even for research purposes.

Beach seining can be useful for **monitoring** annual variations in abundance. All seines must then have the same construction and the sampling procedure must be standardized and constant from year to year. The best beach seine data series for ecological research and monitoring is the "Flødevigen-series", which has sampled 100 fixed stations during September to October since 1919 (Stenseth *et al.*, 1999; Smith *et al.*, 2002). Since 1919, the seine construction, sampling area, and sampling procedure has been standardized.

3.1.2.3 Plankton Nets
Tone Falkenhaug, Institute of Marine Research, Norway

Plankton includes organisms covering a wide range of sizes and many different taxonomic groups. There is thus no single sampling gear that is "ideal" for sampling the whole plankton community. The choice of gear depends on which plankton organisms are to be studied and on available resources. Attempts should, however, be made to limit the number of different gear in order to improve standardization and to enable greater comparability of different studies.

Ring Nets

Plankton ring nets consist of a fine-meshed net attached to a metallic ring that allows capture of differently sized plankton simply by changing the mesh size of the net. At the end of the net is a collection cylinder called a codend. Plankton nets can be used for vertical, horizontal, or oblique sampling, and can incorporate opening and closing mechanisms to sample at desired depths without contamination.

Table 3.1 Dimensions of different plankton nets. Length of forward cylindrical part (h1) and conical after part (h2). See Figure 3.13 for schematic showing cylindrical and conical sections.

	Phyto-plankton net	Juday	WP-2	WP-3	Egg-net
Mouth diameter (m)	0.36	0.36	0.56	1.13	1.60
Mouth opening (m²)	0.1	0.1	0.25	1.00	2.00
Mesh size (μm)	10	180 / 90	180	1000	375
Hauling speed (m/sec)	0.1	0.5	0.5	0.3	0.5
h1 (cm)	46	0.45	0.70	0.52	2.00
h2 (cm)	90	0.95	1.55	2.00	4.48

The main advantage of ring nets is their ease of use and low cost. Ring nets can be towed from any type and size of vessel, are less sensitive to rough weather, and can be used in shallow waters. An ideal ring net must collect plankton with known filtration efficiency, which is determined by three key net parameters: the net mouth opening, net length, and mesh size. Numerous variants of plankton ring nets have been developed to sample specific size classes of organisms. The ring nets that have been most widely used are presented in Table 3.1.

Phytoplankton Nets

A phytoplankton net is a generic term for a sampling net having a mesh size of 150 μm or less that is used to collect phytoplankton. The phytoplankton ring net (Figure 3.12) has a mesh size of 10 μm, a mouth opening diameter of 36 cm (0.1 m²), and a conical net bag of 90 cm length. The net is equipped with a codend that has gauze covered side windows that act as a sieve (allow water to pass out of the codend while retaining organisms. The net is hauled vertically from 50 m to the surface at $0.1\,\mathrm{m\,sec^{-1}}$.

Zooplankton Nets

The **WP-2 net** is a simple ring net with a cylindrical upper part and a conical lower part (Figure 3.13). The WP-2 net is designed to sample planktonic organisms of sizes 0.2–10 mm from the upper 200 m of the water column. The WP-2 net is named after "the Working Party no. 2"; a team of scientists who designed the net in an attempt to standardize field equipment and best practices for a specific size class of organisms (Tranter and Fraser, 1968). Their recommendations are still valid today and the WP-2 net is the most common plankton net in use for standard zooplankton sampling.

The standard WP-2 net has a diameter of 0.57 m (0.25 m²) mouth opening and mesh size of 200 μm. The net is equipped with a removable codend with filtering panels so that the sample can be concentrated effectively during sampling. It is of utmost importance that the mesh size of the filtering panels is similar or less than the gauze of the plankton net (200 μm). The end of the net is attached to a weight (15–20 kg) to aid in sinking. The WP-2 net can also be equipped with a Nansen bottle-type closing device to obtain samples from discrete depths.

Figure 3.12 Retrieval of a plankton ring net. *Source:* G. Sætra, Institute of Marine Research, Norway.

(a)

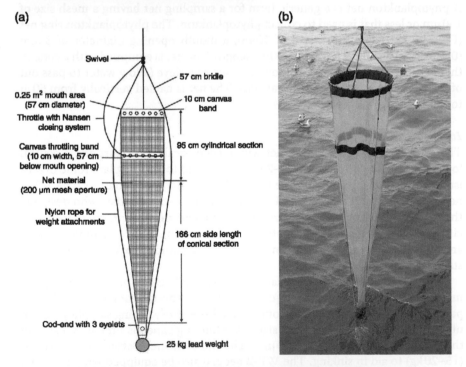

Swivel

57 cm bridle

0.25 m² mouth area
(57 cm diameter)

10 cm canvas band

Throttle with Nansen closing system

95 cm cylindrical section

Canvas throttling band
(10 cm width, 57 cm below mouth opening)

Net material
(200 μm mesh aperture)

Nylon rope for weight attachments

166 cm side length of conical section

Cod-end with 3 eyelets

25 kg lead weight

(b)

Figure 3.13 (a) Schematic of a WP-2 cylindrical-conical net (mesh size 200 μm, mouth opening 0.25 m²). *Source:* Modified from Tranter and Fraser (1968) by R. Jakobsen. (b) Deployment of the WP-2. *Source:* Institute of Marine Research Norway.

For vertical sampling, the net is lowered to the desired depth at $1\,\mathrm{m\,sec^{-1}}$, with care to keep the net off the bottom; a depth sensor attached to the ring can assist with this. The net is hauled vertically at a steady speed of $0.5\,\mathrm{m\,sec^{-1}}$. The net must remain vertical during hauling, that is, with a wire angle of less than 30°. Usually the volume of water filtered is calculated by multiplying the distance integrated by the area of the mouth of the opening. If the net is hauled too quickly or if the net is clogged (e.g. by phytoplankton), the resistance of the net will cause a smaller volume to be filtered. The net can be equipped with a digital flow meter for more accurate volume measurements. This flow meter should not be located in the center of the net mouth opening, but in a position about halfway from the net mouth to avoid overestimating the flow. After retrieval on deck, the content of the net is rinsed into the codend with the seawater hose. The codend containing the sample is removed from the net and brought into the lab for further processing.

The **WP-3** plankton net is similar to the WP-2 net in construction, except the net is larger; the net was designed to sample large sized plankton (>1 mm). The opening has a diameter of 113 cm ($1\,\mathrm{m^2}$) and the conical net bag has a length of 200 cm. Standard mesh size is 1000 μm. Due to the large size, a heavy weight (20–25 kg) and slow towing speed ($0.3\,\mathrm{m\,sec^{-1}}$) is required. The net is usually equipped with a non-filtering codend (0.3–1 liters) for capturing live and delicate zooplankton. After retrieval, the rinsing of the net should be gentle to avoid damaging fragile organisms.

3.1.2.4 Multiple Nets

Multiple net samplers consist of several single nets, which can be opened and closed sequentially on command from the ship or automatically at pre-programmed depth intervals. With nets that open and close on command and a real-time data feed, multiple net samplers provide depth-stratified samples and more precise information on the vertical distribution of zooplankton compared to traditional ring nets. In addition, the ability to put a large number of nets in the water at the same time saves ship time (and therefore resources). However, these instruments are more expensive than ring nets and require both technical competence and a large ship with winches. The multiple net samplers can only sample in open water and are more sensitive to rough weather than ring nets. Collecting plankton close to the ocean bottom or from shallow waters requires different methods.

There are a number of commonly used multiple net samplers that use square mouth-opening nets and come in a variety of sizes. Here, we present the two most commonly used.

The **MOCNESS** (Multiple Opening/Closing Net and Environmental Sensing System; Wiebe *et al.*, 1985) is a computer-controlled multi-net system that enables collection of samples at specific depth intervals in the water column (Figure 3.14, Figure 3.15). The MOCNESS consists of a set of nets stacked on top of each other in a single rectangular frame that is towed behind a ship.

The dimension of the rectangular opening varies, but is chosen so that a precise area of water is sampled. The nets are dyed dark blue or black to reduce contrast with the background (and thus organism avoidance). Each net is terminated with a $1\,\mathrm{dm^3}$ PVC codend, which holds the samples after the nets have filtered plankton from the water. The nets are opened and closed sequentially by

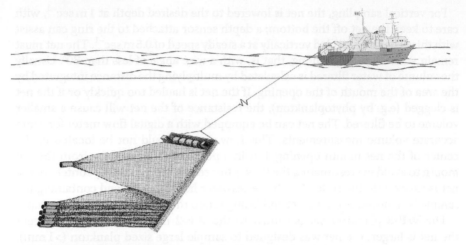

Figure 3.14 MOCNESS in operation. The net is towed obliquely behind the ship at a 45° angle. In this schematic, the eight nets at the bottom have all collected samples at various depths and are now closed; the top net is sampling the shallowest depths on its way back to the surface. The codends at the ends of the nets hold the samples after the nets have filtered plankton from the water. *Source:* Artwork by R. Jakobsen.

Figure 3.15 Deployment of the MOCNESS. The first net (net 0) is kept open while lowering to maximum fishing depth. When the maximum sampling depth is reached, net 0 is triggered to close, and net 1 will open for the ascent. *Source:* K. Mæstad, Institute of Marine Research, Norway.

commands through a conductor cable from the surface. There are several sensor instruments mounted to the frame of the MOCNESS, providing real-time data on depth, frame angle, flow, and net number, in addition to environmental parameters, such as temperature and salinity.

Different versions of MOCNESS are designed to capture different size ranges of zooplankton and micronekton. The MOCNESS-1 (mouth opening $1\,m^2$) carries nine nets of 180, 210, or 333 μm mesh. The MOCNESS-10 (mouth opening $10\,m^2$) is equipped with 5 nets of 500–2000 μm mesh and is intended for capturing macroplankton and micronekton (2–20 cm). For standard sampling of mesozooplankton (size 0.2–20 mm), the MOCNESS-1 ($1\,m^2$; 180 μm mesh) is used.

The successive depth strata are chosen by pre-evaluation of the local hydrography or acoustical observations. In monitoring studies, standard preset depth strata are used, with more narrow depth strata in upper layers (e.g., 0–25; 25–50; 50–100; 100–200; 200–300 m).

At deployment, the first net (net 0) will remain open until the maximum sampling depth is reached, at which point net 0 will be triggered to close and net 1 will open for ascent. During ascent, the MOCNESS should be towed obliquely, with a 45° angle, at a speed of approximately 2 knots, which is obtained by careful monitoring of the winch hauling rate and ship speed. During hauling, the nets are opened and closed at desired depths. When the MOCNESS is retrieved on board, the codends are handled separately for sorting and preserving.

The **MultiNet** (Multi Plankton Sampler; Weikert and John, 1981) is a square mouth opening–closing net, consisting of a steel frame with several net bags that can be opened and closed in sequence from a deck command unit via a conducting cable or from a self-contained, programmable unit (Figure 3.16). The MultiNet is available in three sizes: mini ($0.125\,m^2$, 5 nets), midi ($0.25\,m^2$, 5 nets) and maxi ($0.5\,m^2$, 9 nets). The MultiNet can be used for vertical or horizontal hauls. For horizontal collections, a V-fin depth depressor is attached to the MultiNet to keep the instrument stable and properly oriented.

3.1.2.5 Sledges and Dredges

Sledges (or sleds) and dredges are towed along the seabed capturing organisms living both on and in the substrate. The gear used depends on the size of the organisms, the part of the seabed to be sampled (with many types sampling the upper part of the benthic infauna), and the substrate type. A large number of organisms can be collected with each sample, but these gears have several drawbacks: they can be quite damaging to benthic communities; obtaining a picture of the area sampled is not possible unless the equipment is fitted with video equipment; each gear is selective toward certain organisms, therefore selection bias exists; specimens are frequently damaged during collection; and data are qualitative or, at most, semi-quantitative. Nevertheless, towed gear remains indispensable and is some of the most frequently used type of equipment for surveys with a benthic sampling component.

3.1.2.5.1 *Epibenthic Sledges*

Many designs of epibenthic sledges exist (Figure 3.17). They vary from exceedingly simple (e.g., a metal frame with a collecting net) to complex, for example, multiple nets to allow sampling various levels of the substrate

(a)   (b)

Figure 3.16 Examples of (a) a 0.125 m², 5 nets MultiNet. *Source:* Artwork by R. Jakobsen. (b) a 0.5 m², 9 nets. *Source:* R.D.M. Nash, Institute of Marine Research, Norway.

within one haul. Those with simple designs have the advantage of being easy to repair while at sea or can sample the same, regardless of how they land on the seafloor. Some sledges are able to only sample from level sea floors and soft substrate, whereas others are slightly more robust and can be used on hard substrate or rough ground. Selectivity of the different sledges varies. Some employ chains to stir up the top of the sediment so that organisms enter the net (e.g., the Ockelmann sledge; Ockelmann, 1964).

The **Agassiz trawl,** which is actually a sledge, is towed along the bottom and consists of a simple metal frame with a net (Figure 3.18). The sledge is named after the nineteenth-century Swiss-American scientist Alexander Agassiz. Due to its simplicity, it has three major advantages: the trawl can be deployed to any depth with little previous information about the seafloor; no difference exists between the upper and the lower side of the instrument (i.e., sampling success is independent of which side falls on seabed); and the trawl is robust and can withstand rough physical treatment. Despite the simple design, this is still one of the most-used pieces of benthic sampling equipment.

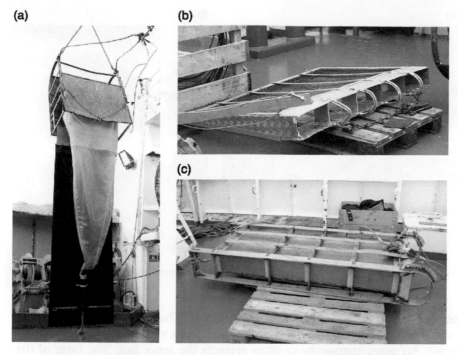

Figure 3.17 (a) RP sledge; (b) Schander sledge; (c) Sneli sledge. *Source:* H. Glenner.

Figure 3.18 An Agassiz trawl on deck. *Source:* H. Glenner.

3.1.2.5.2 Dredges

Dredges are gears for sampling rougher seabeds and harder sediments. Because of their damaging effect on the seabed, their use should be restricted to hard and/or uneven sea floors, unsuited for e.g. beam trawl. A great diversity of dredge designs has been developed for a wide range of such rough substrata (sand to

(a) (b)

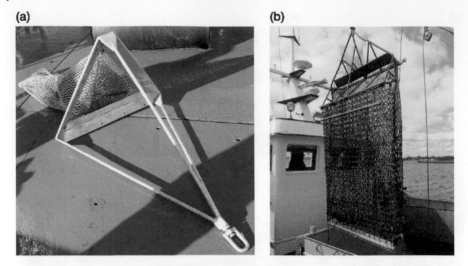

Figure 3.19 (a) Triangular dredge, (b) Blue-mussel dredge. *Source:* H. Glenner.

cobble), but all share the same basic construction. They are simple, solidly build, consist of a robust metal frame and a bag (mesh size varies with targeted organisms). An outer, coarser mesh often protects the inner bag. The form of the dredge depends on sediment and collection purpose, but a common shape is the triangular dredge (Figure 3.19).

A dredge that can be operated over very coarse terrains, is the **blue-mussel dredge** (Figure 3.19). This dredge is also used commercially. The dredge consists of a steel frame, modified to prevent it from digging deeply into the sediment, chain mail, and a solid metal enforced net. At the opposite end of the spectrum is the **Sherman epibenthic sledge**, which could easily also be called a dredge, that was designed to sample invertebrates from rough terrain on seamounts.

3.1.2.6 Grabs and Corers

Grabs and corers allow for quantitative sampling of soft-bottom substrate. These gears remove a section of the substrate and all organisms that live within it. Sample volume (post-collection) is calculated using the amount of free space left in the grab. The sediment sample is then carefully sieved into size fractions using successive sieve sizes (e.g., 5 mm, then 1 mm mesh). To transform the collected and sorted material into data that can provide a meaningful analysis is extremely labor intensive. A single grab sample can contain hundreds to thousands of specimens and requires substantial faunistic expertise. Identification can be very difficult because the organisms are small and differentiation may depend on, for example, differences in mandible structures or number of pores on a leg segment.

Grabs are gear that, with a scoop, collect a sample of the sediment approximately covering the area of the opening (Figure 3.20). The jaws of the grab are open at deployment. The construction of the grab is such that the grab remains

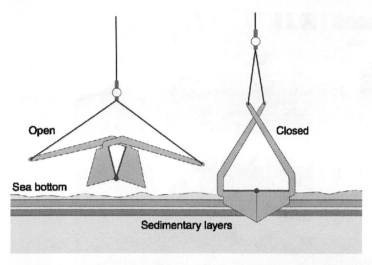

Figure 3.20 Sampling of the sediments using a grab. *Source:* R. Jakobsen.

open and correctly orientated until it reaches the sediment. The design also keeps the gear from over-penetrating the sediment. The jaws either close upon contact with the sediment or when the cable become taut. Grabs are able to collect sediment with very little disturbance (the sample remains as it was in the sediment). The tension on the cable keeps the jaws shut as the gear is hauled to the surface.

Corers work by pushing a tube into soft-bottom benthic substrates bringing on deck intact sediment in the tube/corer. The corer is forced into the sediment either by the weight of the gear or by hydraulics. At the sea floor the lower part of the corer is automatically seal by a cab, which protects the sample against distortion. Sediment cores can be stored chemically, frozen, or be immediately analyzed. Organism group and sediment category determines the type of corer to apply. **Gravity corers** are typically used for collecting the smaller marine metazoans (meiofauna) from the sea floor. **Box corers** take square samples of the sediment. A **multicorer** constitutes of a number of separated individual tubes, allowing for discrete samples to be taken from a single sampling station (typically taken for different, but parallel running analyses; Figure 3.21).

3.1.2.7 Water Samplers
Svein Rune Erga, University of Bergen

The record of creative devices for collecting water from depths of the sea goes back at least to 1666, when Robert Hook designed an instrument for this purpose (Venrick, 1978). Water bottles of various designs were later introduced around the world, however, not all of them were reliable. With the invention of the Nansen bottle/sampler in 1894 by the Norwegian explorer and oceanographer Fridtjof Nansen, reliable samplings at discrete depths downwards in the water

Figure 3.21 A multicorer on deck. *Source:* H. Glenner.

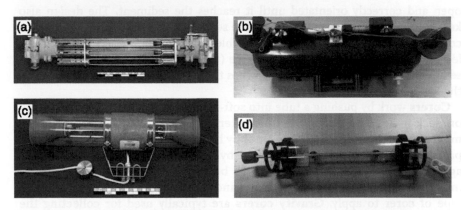

Figure 3.22 Various water samples. (a) The original Nansen sampler with housing for thermometers. (b) The Van Dorn sampler with messenger weight. (c) The newer generation Niskin bottle, shown with lids open, with housing for thermometers on the backside. (d) Ruttner sampler in open position. *Source:* T. Sørlie, University of Bergen, Norway.

column could finally be conducted (Figure 3.22). The water bottles currently most used on research cruises are a newer generation Niskin bottle, introduced in 1966 (Figure 3.22).

Both the new Nansen and Niskin bottles consist of a plastic cylinder with lids that are closed when at the sampling depth, but the closing mechanisms operate differently for each bottle. The bottles are mounted, with both ends open, on a cable and lowered to the desired depth by a winch. A **messenger** weight is dropped down the cable and when it hits the closing device, the water sampler closes. The Nansen sampler closes by turning the cylinder upside down (the reversing principle). The mechanism on the Niskin closes the bottom and top lids by releasing a tension device (either a string or a rubber connecting the caps), which keeps the lids in the open position while being lowered; known as the non-reversing principle for closing. The Van Dorn bottle, introduced by William G. Van Dorn in 1956, also uses the non-reversing principle (Figure 3.22).

The Nansen, Van Dorn, and Niskin bottles can be used for vertical sampling at discrete depths within one cast. Multiple bottles or samplers are mounted on the cable, with messenger weights mounted in between. When the bottles are at the appropriate sampling depths, the messenger is released for the topmost bottle, which, when hit, will release the next messenger (and so on).

A smaller sampler, easily handled from small boats, is the Ruttner sampler (Figure 3.22; Ruttner, 1963). The bottles are Plexiglas cylinders with lids at both ends. It operates similarly to the Niskin bottle.

A Multi Water Sampler system (i.e., rosette or carousel) is normally preferred on research vessels because it can mount many bottles of various sizes for comprehensive sampling of the water column (Figure 3.23). Water bottles are attached to a circular frame, equipped with an electrical release mechanism connected to a depth meter; a CTD is usually attached to the bottom of the ring. The water samplers are closed by a signal from the computer or command unit on deck when the sampling depth is reached.

An important consideration, regardless of the sampler method used, should be the material used for the water bottles. Sterile samples are sometimes needed when collecting microorganisms. Alternatively, pollution can be a problem when water chemistry sampling is needed. Therefore, the material must be autoclavable.

Figure 3.23 An example of a Multi Water Sampler System, with attached CTD. *Source:* Institute of Marine Research, Norway.

3.1.3 Passive Gears

Passive gears are set, either on the bottom or in the water column, and organisms are caught as they come into contact with the gears. Gillnets and entangling nets, traps, and fyke nets are examples of passive gears.

3.1.3.1 Gillnets and Entangling Nets

Gillnets and entangling nets are panels of single, double, or triple netting walls, set near the surface (driftnets), in midwater, or at the bottom (Figure 3.24). Driftnets are typically used to catch schooling species, for example, herring, mackerel, tuna, salmon, squid. Their use is illegal within EU waters and many other areas because of their indiscriminate capture, including sea turtles, seals, and cetaceans. Nets set on the bottom target bottom species, such as cod, ling, or monkfish. The nets have small, solid floats on the upper line (headrope) and weights on the bottom line (footrope), which are distributed evenly along the net. The balance between the weights and the floats determine where in the water column the nets can be positioned. Driftnets drift freely with the current, connected to the operating vessel or a buoy, whereas the bottom or midwater nets are stationary (attached to the bottom).

Gillnets can be hauled by hand, at least for small-scale sampling and at shallow depths if the total length of the net is not too long. Nets that are long or set at depth require more than human strength to haul; a net drum operated by a winch or a power block are commonly used when hauling nets. In small-scale sampling, gillnets are most often folded on deck within a delimited area (to avoid tangling) or rolled up on net drums.

Fish are captured in nets by several methods. Fish are gilled when the head goes through the mesh and the mesh becomes stuck behind the operculum. Fish are wedged when the fish attempts to swim through the net but becomes held tightly around the body by the mesh. Fish are entangled either in a pocket made of net or by being caught by teeth, maxillaries, spines, or other projections.

Gillnets and entangling nets are **selective** gears. Gillnets are made up of meshes of a particular size, which target fish of a particular size (larger fish are not likely to be entrapped, smaller fish may swim through the mesh). Furthermore, where the nets are set will target a particular species (or multiple species). For a particular mesh size and target species, the captured fish have a high probability of belonging to a certain length class with decreasing probability of capture of larger and smaller length groups. No gear exhibits equal selectivity toward fish of all sizes within a population. This means that selectivity may affect any estimates that require random sampling from a fish population. It is essential to consider gear selectivity prior to sampling fish for quantitative measurements (for details, see Cochran, 1977; Gunderson, 1993; Engås and Løkkeborg, 1994). Examples of application of theories of selectivity on datasets of cod are illustrated in Salvanes (1991). The most important factors in selectivity are mesh size, net construction, the shape and behavior of target species, and method of capture (e.g., by entanglement, by gilling; Hamley, 1975).

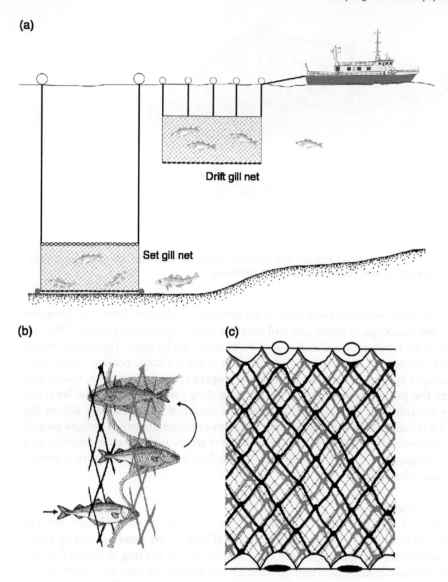

Figure 3.24 Schematic showing gillnets and entangling nets. (a) Gillnets can be set at the bottom or midwater, where weights and floats keep the net open, upright, and at the correct depth; buoys at the surface are then used to find the nets for retrieval. Drift nets are set at the surface. (b) Entangling nets operate by capturing fish as they attempt to swim through the net. (c) Schematic showing the weights along the groundrope and floats along the headline, which keep the net fully open when set. *Source:* Artwork by R. Jakobsen.

3.1.3.2 Pots

Pots are cages or baskets that range in size from small bottles up to two meters, typically set on the bottom, in a wide range of depths and habitat types (Figure 3.25). Some are set on smooth, muddy, or sandy bottom, while others are set close to rocks, coral reefs, and wrecks. They are made from various materials,

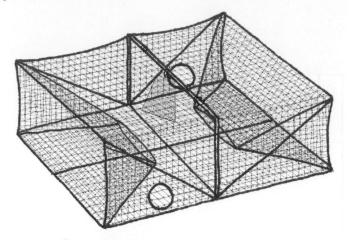

Figure 3.25 An example of a lobster and crab pot. The open holes illustrate escape openings for undersized lobster. *Source:* Artwork by R. Jakobsen.

such as wood, wicker, metal rods, wire, netting, or plastics. Pots may have one or more openings or entrances and may use bait. Commonly, pieces of fish are used as bait, but artificial or flavored baits may also be used. Typically, the targeted species are crustaceans (e.g., lobster, crab, shrimps, scallop, *Nephrops*), squids, or species of fish that tend to congregate around structures. Catch may enter the pot, either for shelter or because they are attracted to the bait, but they cannot exit. The size of the mesh or distance between slats allows the smallest individuals to escape; some fisheries regulate the use of **escape panels** for undersized individuals. Pots are often set singly or in rows connected to a line. Single pots can be operated by hand, but pots in row require a **power block** for hauling.

3.1.3.3 Fyke Nets
A fyke net is a fish trap, consisting of a cylindrical or cone-shaped net bag mounted onto rings or other rigid structure (Figure 3.26). Leading nets on either side of the net bag guide fish into the entrance of the net bag. A single fyke net may have two leading nets and be fixed on the bottom or have double entrances facing toward each other, with one leading net in between (Figure 3.26). Fish enter the fyke net, often through a funnel structure, but cannot exit.

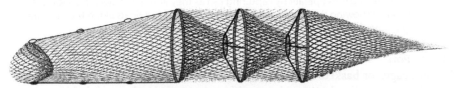

Figure 3.26 An example of a cod fyke net. The leading net is to the left and the net bag with funnels is to the right. *Source:* Artwork by R. Jakobsen.

3.1.3.4 Hook-and-line
Keno Ferter, Institute of Marine Research, Norway

A large range of hook-and-line gears exist. **Longlines,** a passive gear, are set vertically or horizontally and have at least one end of the line attached to a buoy at the surface for easy retrieval. Anchors are often used to fix longlines on the bottom to avoid drifting. Other hook-and-line gears are fished actively and tend to be fixed to the boat, shore, or held by a person. Angling is included because large predators can often evade capture by the trawl. In some studies, angling may provide better information on the vertical position of the fish compared to other methods. For diet studies, angling for samples may give better results because regurgitation may occur less often. Angling, however, will capture only those individuals that are actively searching for prey.

Hook-and-line gear consists of different components: the terminal tackle, the fishing line, and the rod and/or reel for angling. Terminal tackle describes everything that is fastened to the end of the line. This includes, but is not limited to, floats, weights, swivels, hooks, and lures. The type of tackle one will use for sampling depends on many factors, of which the most important are the target species, the area where one wants to fish, and the weather conditions.

To attract the fish to ingest the hook, artificial lures, natural bait, or a combination of these can be used. The size of the lure or bait is determined by the target species and its size. If one wants to catch smaller individuals, smaller lures and baits should be used, while larger lures and baits generally select for larger individuals. The successful hooking of a target species depends on the type of hook used. Moreover, hook type influences hooking injuries and hooking location, which play an important role if one wants to keep the fish alive after capture. Swivels minimize twisting of the line and weights act to sink the lure or bait to greater depths. Data storage tags fixed to the terminal tackle can be used to determine capture depth and temperature.

3.1.4   Remote Sensing

Remote sensing is a method to obtain information about objects or areas from a distance, where the observing sensors are not in direct contact with the measured item. Here, we detail a few methods of remote sensing.

3.1.4.1   Acoustics
Egil Ona, Institute of Marine Research, Norway

Sound has been used to locate and visualize organism distribution, abundance, and behavior for nearly a century (Horne, 2000; Simmonds and MacLennan, 2005). Originally, scientists and commercial fishers used sound to locate dense aggregations, but advances and developments soon resulted in the ability to make quantitative measurements of abundance. These advances also meant that more intricate studies of fish behavior using acoustics were also possible. Single organisms could be tracked through the water column, but these types of observations were more reliant on stable platforms, not moving vessels. Typically,

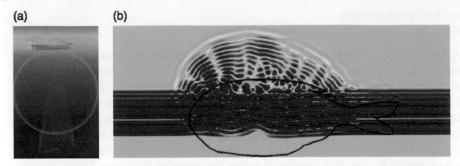

Figure 3.27 Schematic illustration of how an echo sounder works. (a) sound is directed downwards from a transducer located on the hull of a vessel, some of which is reflected back to the transducer by organisms, such as fish schools in the water column or located near the water bottom, as well as the seabed. *Source:* Institute of Marine Research. (b) Diagram of how sound is reflected from a fish. *Source:* G. Macaulay, Institute of Marine Research, Norway.

behavioral observations are made from stationary vessels or platforms, or with mounted transducers, resting either on the seafloor or moored at various depths in the water column.

Acoustics is the sending of waves of sound energy through the water from an acoustic transducer; these sound waves can be thought of as a pressure wave. The transducer is capable of both transmitting and receiving sound. The transmitted sound is reflected when it hits an object and some of this reflection will be back in the direction of the transducer (Figure 3.27). Echoes, or reflected sound, returning to the transducer are called backscattered sound. Anything hard or having a density contrast with the surrounding seawater, positive or negative, will reflect sound, for example, rock, shell, or air, such as that found inside the swimbladder of a fish (see Simmonds and MacLennan (2005) for a full description of modern fishery acoustics). The larger (or denser, or lighter) the object, the greater the amount of sound that is reflected back. The ability of an object to reflect sound is described through its target strength. For example, large swim-bladdered fish tend to reflect more sound than fish without a swimbladder, and large fish more than small fish. Organisms that possess gas, having either a gas-filled swimbladder or gas bubbles (some plankton or zooplankton), will reflect more sound than organisms that do not possess gas. Similarly, organisms with fat will generate a stronger echo than a similar-sized organism without fat.

The seafloor is a huge, more or less flat target for the acoustic wave. The echo is usually 10 000 (40 dB) stronger than any of the targets in midwater and can therefore easily be separated from other targets. In this section, we will describe how to use acoustics, echo sounders, and sonar for fishery and marine biology investigations. Acoustic methods for bottom mapping and habitat mapping will only be briefly mentioned here.

The frequency of the transmitted sound determines the range at which the echoes from the preferred targets can be detected. This is due to the physical attenuation of sound, which is frequency dependent, just as is the attenuation of

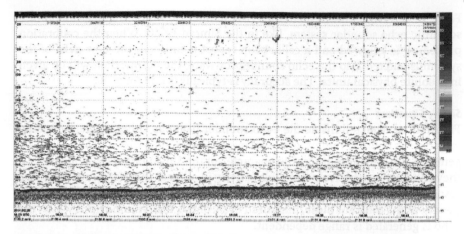

Figure 3.28 Echogram from Nordkapp of predominantly single targets at relatively low density, where every horizontal mark is an individual fish. The individual marks within the bottom 150 m are Atlantic cod. The small schools near the surface are capelin and are made up of many individuals. *Source:* E. Ona.

light by water. A frequency of about 30 to 40 kHz was discovered to be suitable for fisheries investigations. At these particular frequencies, transducers enabling the sound beam to be concentrated within a 10 degree beam could also be made at reasonable cost. A 38 kHz split beam transducer with an opening angle of about 7 degrees can detect a single large cod to about 900 m depth and an echo from the seabed to about 3000 m. The effective detection range for smaller targets, like a single myctophid fish, is about 500 m; the detection range is very much dependent on the noise level of the vessel and machinery at this frequency. This noise can be recorded at the transducer with the echo sounder running in passive mode. Both research vessels and larger fishing vessels carry echo sounders of one or several different frequencies.

Backscattered sound is visualized in a two dimensional echogram (Figure 3.28). The x-axis on the echogram normally displays time or distance (e.g., when recorded from a moving vessel), while the y-axis shows depth below the transducer. Colors display the intensity of the reflection; typically the stronger echoes appear red, but this will vary depending on the color scale chosen for visualization. Echograms, however, only provide part of the sample; they show the location of organisms, but give no indication of the species or size ranges. To identify and verify what is seen on the echogram, target identification is needed. This is typically done by sampling organisms by trawling, camera drops, or other sampling techniques.

Calibration

Fishery acoustics use instruments and digital output that should be calibrated according to a physical standard. Echo sounders are routinely (several times per year) calibrated using a reference target positioned under the vessel inside the center of the acoustic beam. Procedures for this calibration are found in

Foote *et al.* (1987) and Demer *et al.* (2015), but also inside the help menu of a scientific echo sounder system, like in the Simrad EK60. When the calibration procedure is followed and a calibration report is finished, the new **gain** of the system is entered into the echo sounder software; the output then will be given in absolute physical units. The accuracy is typically 0.1 dB, or ±1%.

Simple Situations

Fish may be detected inside the echo sounder beam as either single, separated targets or as multiple targets. If several targets inside the same echo pulse volume generate the echo amplitude, the character of the single targets disappears, but they add up constructively and destructively in a linear manner. Since the pulse resolution is defined by the area of the beam at any range and the pulse duration (length of the transmit pulse, measured in seconds), unit volume from which the echo is generated is range dependent.

If we use a transducer with a 7 degree half power width, the so-called equivalent (ideal) beam angle (see Simmonds and MacLennan, 2005) covers about 0.00845 sr (steradians), which is the fraction of the surface of a sphere covered by this angle. In a sphere, there are 4π steradians.

The effective area of the acoustic beam at 100 m range is therefore

$$A = \psi r^2 = 0.00845 * \left(10000\right) = 84.5\,\mathrm{m}^2. \tag{3.1}$$

If the pulse duration is $1\,\mathrm{ms} = 1/1000\,\mathrm{s}$, the length of the echo is

$$c * \tau/2 = 1470 * 0.001/2 = 0.735\,\mathrm{m}, \tag{3.2}$$

and the resolution volume is

$$84.5 * 0.735 = 62.1\,\mathrm{m}^3. \tag{3.3}$$

At this range, if the target densities are low, we will obtain single target recordings, like those exemplified in Figure 3.28.

The representation of each target is different because the number of detections of the same target when the vessel passes over a stationary target will depend on vessel speed, ping rate, and the width of the acoustic beam. A time-varied-gain function (TVG) inside the echo sounder will compensate for transmission loss and ensure that the target amplitude is independent of depth. If the vessel passes straight over the target, the target will enter the outskirts of the beam in the first detection, then gradually move towards the center of the beam, and finally pass out on the aft part of the beam, forming an **echotrace** that resembles an inverted V on the echogram. The range to the first and last detection is larger than when it is the center of the beam, explaining the V-shape of the trace in the echogram. Wider beams will typically give extended V-shapes of such targets.

In some surveys, the number of targets can be counted over a specified distance; from computing the effective observation volume of the beam, the number of fish per nautical mile squared can be estimated. In special circumstances, a full density estimation by target counting can be completed. Modern echo sounders are now directly connected to GPS systems and will mark each nautical mile (mile marker). Transects crossing the registration are then used to generate the density in each strata in a typical survey.

Fortunately, in 1965, the echo integrator was invented (Dragesund and Olsen, 1965) and total echo energy of echoes in a defined layer on the echogram could now be integrated (summed). This echo integral was later proven to be directly proportional to fish density, and could, more or less, be used to directly compute fish density. The only unknown parameter was now the mean target strength of the fish surveyed.

If ρ_A is the area density of fish (n/nmi^2), s_A is the area scattering coefficient, and σ is the mean backscattering cross-section (or the linear version of the Target Strength; TS), then the density of fish can be estimated as:

$$\rho_A = \frac{S_A}{\sigma}. \tag{3.4}$$

In the case of the previous example, both the mean area scattering coefficient and the mean backscattering cross-section can be measured with the same echo sounder. The split beam principle, where the target position within the beam can be determined for a single target, was introduced in the 1980s (Ehrenberg, 1979; Brede *et al.*, 1990). The problematic beam compensation of the echo strength was then solved and the target strength of the object itself could be measured directly. With a properly calibrated echo sounder (Foote *et al.*, 1987), both the nominator and the denominator of the equation could be measured in absolute physical units. In the cod example (see Figure 3.28), the echo integral or area backscattering coefficient for the nautical mile shown is 1173 (m^2/nmi^2) and the mean target strength is measured by split beam to be close to −25 dB (−24.94 dB), ($<\sigma> \approx 0.04025$ m^2) when averaged in the linear domain. The density measured is close to the counted density, 29140 fish/nmi^2 or 0.00084 fish/m^3.

When the targets are more concentrated in a simple monospecies aggregation, as in schools or layers, the same formulae apply. The next example shown is from a herring aggregation in Ofotfjord, where a large fraction of the Norwegian spring spawning stock wintered in the period 1987 to 2003 (Figure 3.29). A survey with transects crossed the fjord in a systematic manner, using 1 to 5 nmi spacing, and surveyed the entire herring population. The mean density for each nautical mile was computed from the area scattering coefficient. As shown in Figure 3.29, mean density was 53000 m^2/nmi^2 for mile 4. Using the target strength relationship: TS = 20logL −71.9 dB (based on the recommendations of Foote (1987), estimate of a mean TS for a 32 cm herring was −41.79 dB (or 0.0083 for mean σ). This would then give 6.38e^7 herring per nmi^2, or 18.6 herring per m^2 for this particular mile. The registration is not resolved, meaning that all the herring are registered as multiple targets.

Scrutinizing Echograms

From the previous examples, we see that we can measure single and multiple targets in clean concentrations. Trawling with large pelagic trawls are used to sample pelagic target species, while demersal trawls may be used to sample

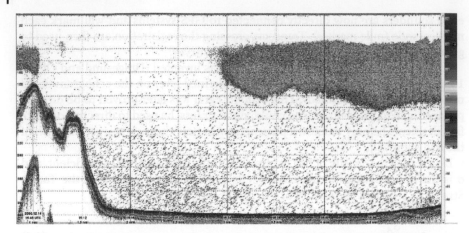

Figure 3.29 Example echogram of herring in Ofotfjord. The green to red colored targets from 50 to 150 m depth are herring and the faint (blue) single targets at deeper depths are blue whiting. The echogram shows five nautical miles (9.3 km), where individual miles are marked with vertical lines. After safely removing the bottom echo (red line), the echo integral for mile no. 4 is 53 016 (including the blue whiting) and 53 000 without the blue whiting.
Source: Institute of Marine Research, Norway.

targets close to the bottom. Trawling has two purposes in acoustic surveys: to identify species in the registration and to obtain biological samples and size distributions. In a particular survey, the main target category is selected (e.g., herring, cod); echograms are then scrutinized and interpreted with this as the main category. This means, in practice, that other targets (species) are regarded as noise and often left uninterpreted. In the previous example, only herring are interpreted and stored to the database, while cod, blue whiting, and plankton are left unassigned.

The experience of the person scrutinizing the data (interpreter) is very important. Experienced interpreters, who have been observing and sampling a specific ocean, may easily distinguish many target categories in the water column, while inexperienced interpreters may have to trawl frequently to ensure their interpretation is correct. Multifrequency acoustics, where 3–6 frequencies are used simultaneously, can be very important for identification of different target categories (Korneliussen and Ona, 2003; Johnsen *et al.*, 2009). Categories are identified by using the frequency response. Some species and sizes have such clear responses that they may be automatically used to separate the target category from a mixed school situation. Similar approaches are also used for separating zooplankton targets like krill and copepods (Horne, 2000). If a multispecies approach is needed, where interpretation and separation of several target categories on the echograms is needed, then the precision and accuracy of each category will be reduced due to the sampling capacity of the vessel.

Acoustic surveying is usually performed according to a well-established survey design that incorporates some sort of stratification. Because the stock is assumed to be stationary, or approximately stationary, during the survey, completion of the survey must happen quickly to avoid migration issues. If surveys are conducted during migration, then a correction must be applied. See further examples in Simmonds and MacLennan (2005).

The most common errors in acoustic surveying methods are described by Aglen (1994) and include: blind zone close to the sea surface, air bubble attenuation, vessel avoidance, inappropriate stock coverage, interpretation errors, and calibration errors. If the survey is appropriately conducted and an error analysis is made, the total uncertainty in fishery acoustic surveys can typically be less than 30%, with respect to absolute stock abundance. Reducing the uncertainty by using increased effort should be balanced with respect to the error sources mentioned earlier.

The methods described above have also been extensively used to study diel vertical migration (DVM) in, for example, mesopelagic fish or demersal fish. Acoustics are then used from a relatively stationary vessel or from fixed stationary systems that either float at the surface or are mounted on the bottom, pinging towards the surface. Feeding in different trophic levels occurs often during the dark periods of the DVM cycle, and a specific plan for sampling throughout the cycle should be made for understanding the real interaction. Remember that the different sampling gears used have significantly different sampling efficiency across both species and size. If non-closing sampling devices are used, contamination of the sample from the depth may be large during setting and retrieval. Interpretation of the samples with respect to catch efficiency may then be needed. Modern cameras with and without light can also be used to monitor the catch in the codend of trawls in order to better understand when the different organisms entered the trawl, and sophisticated equipment, like the DeepVision (Rosen *et al.*, 2013) can be used to quantify the catch process.

3.1.4.2 AUVs
Lars Asplin, Institute of Marine Research, Norway

A **glider** is an autonomous underwater vehicle (AUV) that uses wings and small changes in buoyancy to move forward in oscillating slow ascent and descent patterns (Figure 3.30). The glider can travel long distances, equipped with a variety of (not too energy demanding) sensors measuring physical or biological parameters, making profiles in this vertical zig-zag pattern. Each time the glider surfaces, it can transmit data, as well as receive instructions on how to operate. Typically, these gliders require a long horizontal pathway and hence are mostly suitable for the oceans.

3.1.4.3 Satellite or Infrared Light
Svein Rune Erga, University of Bergen

Remote sensing can be done using aircraft or satellites to detect reflected or emitted energy from the earth, where sensors respond to external (passive) or internal (active) forces. The amount of electromagnetic radiation (EMR) from the sun, reflected back from the ocean to sensors on satellites, is one type of external stimulus (Figure 3.31).

Figure 3.30 A Kongsberg Seaglider on deck, ready to be launched into the Norwegian Sea. *Source:* T. Hovland, Institute of Marine Research, Norway.

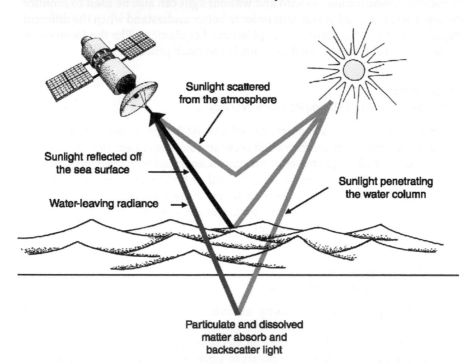

Sunlight scattered
from the atmosphere

Sunlight reflected off
the sea surface

Water-leaving radiance

Sunlight penetrating
the water column

Particulate and dissolved
matter absorb and
backscatter light

Figure 3.31 Main principle of ocean color remote sensing by a satellite. Some part of visible light is scattered by the atmosphere and some is reflected at the sea surface. These signals can be corrected for, thereby allowing the "water-leaving radiance" to be the signal containing information about water constituents. *Source:* Redrawn from Dierssen and Randolph (2013).

The EMR received contains information, often as an image that comes from the ocean or land. Understanding how the atmosphere influences the EMR signal, through absorption or scattering of reflected radiation, needs to be understood to interpret the image correctly. Particulate and dissolved matter in water may also influence the spectral composition of the reflected radiance. Phytoplankton strongly contribute to the spectral change in radiance during growth seasons (e.g., during blooms). Pigment composition, which varies with the type of microalgae, is responsible for absorption in the visible part of the light spectrum, while structural components (e.g., size, shape, type of cell covering) determine how the light scatters. Because of this, each species of microalgae has its own optical signature and can be recognized by the satellite. Ocean color data received by satellites are images, necessitating having specific analyzing and interpreting tools, including the use of computer algorithms. Algorithms have greatly improved over time, giving better tools to detect and monitor phytoplankton blooms (Blondeau-Patissier *et al.*, 2014).

Satellite ocean color sensors have been used since 1978, with the launch of the Coastal Zone Color (CZCS, NASA), for mapping and monitoring marine resources. A number of ocean color sensors were later developed and launched, including SeaWifs (Sea-viewing Wide Field-of-view Sensor, NASA), MODIS (Moderate Resolution Imaging Spectroradiometer on board the Terra and Aqua satellites, NASA; Figure 3.32), and MERIS (Medium resolution Imaging Spectrometer; ESA; Figure 3.33) (Zhao *et al.*, 2014). Improvements in quality have been ongoing and are greatly assisted by comparisons of satellite derived data with *in situ* measured data.

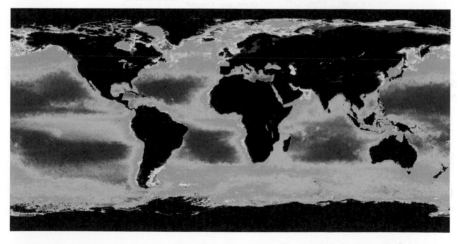

Figure 3.32 A false-color picture of the global annual mean of chlorophyll *a* (mg m^3) based on satellite remote sensing from Aqua-Modis (NASA), Ocean Color web, http://daac.gsfc.nasa.gov./MODIS/

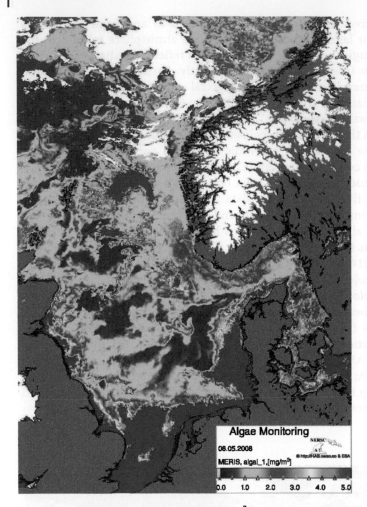

Figure 3.33 Chlorophyll *a* concentration (mg m³) in the North Sea and Skagerrak based on satellite remote sensing measurements from European MERIS (ESA), representing May 8 2008. *Source:* With permission from Nansen Environmental and Remote Sensing Center, Norway, http://HAB.nersc.no

3.2 Sampling the Physical Environment

To draw conclusions about mechanistic relationships or factors influencing species, measuring the physical environment the organisms inhabit is needed.

3.2.1 Conductivity, Salinity, Temperature, Oxygen

3.2.1.1 CTD
A CTD measures conductivity, temperature, and depth (shown on Figure 2.7 and Figure 3.34). Systems can have additional sensors added, including an oximeter,

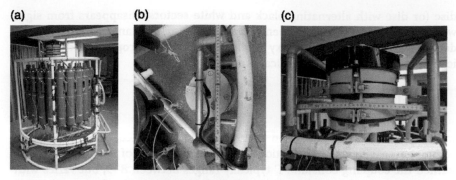

(a) **(b)** **(c)**

Figure 3.34 (a) A CTD rosette with mounted water samplers and a Lowered Acoustic Doppler Current Profiler (LADCP) mounted on top. (b) One LADCP is mounted to face downwards, while (c) the other points upwards. *Source:* M. Dahl.

transmission meter, and fluorometer. CTDs come in a range of sizes, from small, handheld units to large units deployed on research vessels. A CTD takes measurements at predefined depth or time intervals, depending on how it is deployed.

If deployed from a vessel to take profiles of the water column, recordings are generally one per 1 m. The CTD is lowered vertically at a speed of $0.5\,\mathrm{m\,s}^{-1}$; measurements are recorded on both the down- and up-cast. The data are downloaded upon retrieval as (typically) ASCII text files. Salinity, temperature, and oxygen concentration data have to be transformed at download, before they are ready for use.

3.2.1.2 Weather Station
Most ships have a weather station on top of the wheelhouse that automatically logs current weather conditions. The crew may also manually collect information on wind scale, cloud cover, and other weather variables. These data may be stored within the CTD ASCII text file (typically within the first 25 rows of data) or within other vessel log files.

3.2.2 Light

Dag L. Aksnes, University of Bergen

There are many different instruments for measuring light and optical properties of the water. Here, we focus on the secchi disk, transmission meters, PAR sensors, and spectroradiometers.

3.2.2.1 Secchi Disc
Secchi depth observations are recognized as a valuable proxy for detecting long-term changes in the water clarity of oceanic and coastal ecosystems. This is because of the unfortunate situation that long time series obtained by radiometric instruments are lacking. Secchi depth, the depth at which a white

disc (or disc with alternating black and white sectors) disappears from sight when lowered in water, has been measured for more than a century. Secchi depth (S, m) is often used as proxy for phytoplankton (chlorophyll) and eutroph-ication, but it is basically an optical property:

$$S = \Gamma/(K + c) \tag{3.5}$$

where K (m^{-1}) is the attenuation of downwelling irradiance, c (m^{-1}) the beam attenuation coefficient (see Section 1.3.1 for explanation of K and c), and Γ is termed the coupling constant. Γ typically ranges between 8–9, but varies with several factors, such as disk size and color, the observer, and reflectance proper-ties at the surface. A change in secchi depth is commonly reported as the change in meters. Because K and c relate inversely to secchi depth, variations in the reciprocal secchi depths are often of interest. In this context, a one meter reduc-tion in secchi depth (from 2 to 1 m) requires more than 45 times increase in $c + K$ than a one meter reduction from 10 to 9 m secchi depth.

3.2.2.2 Transmission Meters

Light transmission meters are often attached to the CTD unit. Transmission meters consist of an artificial light source and a light sensor, therefore, they are not dependent on daylight and can be used in darkness. The light source emits an artificial light beam at one specific wavelength. Some photons from the beam are absorbed by the water and its constituents, some are scattered out of the beam, and the rest reach the sensor. This means that the instrument measures the fraction of the light beam that remains and, consequently, the beam attenuation coefficient, c (which is the sum of absorp-tion and scattering) can be estimated. If the path length (r) is 25 cm and the measured transmission (T) is 0.9, then the beam attenuation coefficient (c) is: $c = -\ln(T)/r = -\ln 0.9/0.25 = 0.42 \, \mathrm{m}^{-1}$. The value of c, like scattering and absorp-tion, is dependent on wavelength.

3.2.2.3 PAR Sensors

PAR measurements were invented as a proxy for the energy available for pho-tosynthesis and are mostly made for the euphotic zone, that is, the upper 20–100 m depending on water clarity. PAR is the total energy (number of photons) between wavelengths 400 and 700 nm (see Section 1.3.1). PAR instru-ments have become standard on research vessels and are commonly attached to the CTD. Some underwater PAR sensors are spheres and measure the summed radiation/photons that arrive from all directions (**scalar irradiance**). Other sensors are flat. If a flat sensor is facing upwards, which is the normal operational mode, it measures **downwelling irradiance** (see Section 1.3.1). In addition to the underwater sensor, a deck sensor is often installed (commonly at the top of the ship where disturbances by artificial lights are minimized) to continuously measure incoming sunlight. Since sunlight might change (e.g., due to cloudiness) during lowering of the underwater sensor, simultaneous measurements with the deck sensor are useful to standardize the underwater measurements. Older sensors often have a relatively low sensitivity and indicate "zero" light intensity (no light for photosynthesis) when enough light is still present for light sensitive animals (e.g., for vision and behavioral responses in mesopelagic fishes).

3.2.2.4 Spectroradiometers

A spectroradiometer measures electromagnetic radiation in many bins of different wavelengths, including outside the range of visible wavelengths. For example, the RAMSES instrument has sensors that measure light simultaneously in about 200 bins, each around 3 nm, in the band from 280 to 950 nm. Thus, in contrast to the single PAR measurement, 200 irradiance measurements can be obtained for each depth. Hence, 50 different depths result in 10 000 observations from the underwater sensor. Post-processing of the data require software other than a spreadsheet.

3.2.3   Currents (Direction, Speed)

Martin Dahl, Institute of Marine Research, Norway

Measurements of salinity and temperature of fjord water masses are the easiest way to collect information on physical properties and fjord dynamics. However, direct measurement of the current will always be a better option to estimate water exchange and transportation.

Several methods for observing current exist and can roughly be divided into stationary observation platforms or devices that move along with the water. Stationary platforms include traditional current meters with either rotors to measure speed or, more recently, sensors using sound to measure doppler shift and thus water speed. Stationary current meters can measure the current either at the meter itself or along a profile from the current meter to some distance from the instrument (so-called acoustic doppler profilers). These devices also use an embedded compass to estimate directions. Platforms moving along with the water can typically be a drogue with a sail and a GPS that measures the position frequently.

A more expensive way of measuring surface current is to use a land-based radar system or even radars from satellites.

3.2.3.1 ADCP (Acoustic Doppler Current Profiler)

An ADCP measures currents in coastal and open oceans. This is, for example, standard equipment on all research ships of the Institute of Marine Research (IMR) in Norway. ADCPs work by continuously measuring the speed and strength of currents by depth; they quickly and efficiently provide an overview of the current systems. The measurements are provided by transducers (e.g., 75 and 150 kHz) mounted on the ship hull. The measurements are then processed using a computer program (e.g., VMDas), which also controls the storage of the raw

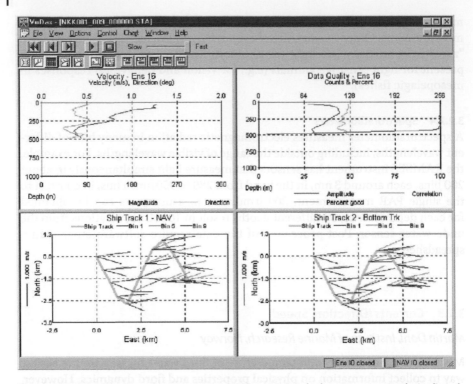

Figure 3.35 Visualization of currents using VMDas. (a) Graph showing magnitude (red) and direction (green) of current as a function of depth. (b) Graph shows data quality as a function of depth. (c) Plot of ship track showing the current velocity vectors along the ship track at three separate depth bins. (d) Same as plot in bottom left, after setting bottom track as the reference velocity; minor deviations in track and current velocities may be apparent because the reference velocities have been subtracted from the profile before output. *Source:* M. Dahl.

data and the ship's navigational data. VMDas visualizes mean current speed (usually every 10 minutes) by plotting current speed by depth, direction, and strength along the vessel track by chosen depth bins, where different bins are visualized by different colors (Figure 3.35). Current directions are presented opposite from how wind direction is measured; a northern current goes toward the north, while a northern wind comes from the north. For example, a current of 180 degrees means that the current flows in the direction of 180 degrees, the current does not come from that position.

Principles and Definitions for ADCP Measures

The acoustic transducers measuring currents utilize the **doppler effect** (change in frequency and wavelength of a soundwave) from particles floating passively in the water masses. To account for variations in the direction and strength of the current, the water column is divided into depth ranges (bins) that each have a

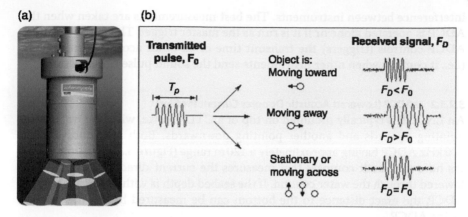

Figure 3.36 (a) A boat mounted ADCP transducer surface with four transducers. The direction of two of the beams are along-ship and two are across-ship. (b) Simple description of the Doppler effect. *Source:* Reprinted with permission from Son Tec.

thickness determined by the entire range of the acoustic current profiler or by the bottom depth. The range depends on the frequency used by the instrument. Generally, low frequencies mean a longer range is possible, while high frequencies have shorter ranges. Typical ranges are:

- 75 kHz circa 700 m
- 150 kHz circa 400 m
- 300 kHz circa 120 m

A ship mounted ADCP consists of four transducers mounted according to the direction of the ship; one faces to the right, a second to the left, a third toward the back, and the fourth faces forwards (Figure 3.36). The ADCP measurements can then be related to the gyro and navigation system on the vessel. A pulse is sent out from each of the transducers (F0). If the received reflected signals (FD) are higher than original (FD > F0), the particles (reflecting objects) and thus currents, are moving toward the transducer (Figure 3.36). If the reflected signal is less than the original pulse (FD < F0), the particles are moving away from the transducer. If FD = F0, the particles are stationary (i.e., there is no current). The time between the transmitted pulse and the received signal provides information on depth or distance, while any change in frequency provides information on speed.

If the seabed bottom is within the range of the ADCP, the instrument can measure true current direction and speed. Otherwise, the ADCP measures current relative to the ship's track.

The measurements taken by the ADCP must be independent from the ship's movements in all directions. ADCPs must therefore connect to high precision reference instruments that provide information on the direction, speed, roll, pitch, and heave of the ship. ADCPs are commonly used at the same time as other acoustic instruments, such as sonar and echo sounders. An inter-calibration between the different instruments must be conducted to avoid interference between instruments. The best measurements are taken when the ADCP is operated alone or if it is run as the master trigger. This means that the ADCP controls (triggers) the transmit time of all other acoustic instruments (i.e., it controls when other instruments send the sound pulse transmission).

3.2.3.2 LADCP (Lowered Acoustic Doppler Current Profiler)

An LADCP is typically mounted on top of a CTD-rosette, with one transducer pointing upwards and another pointing downwards. Both transducers are a 300 kHz ADCP having approximately a 120 m range (Figure 3.34). The transducers have an inbuilt compass that measures the current direction. The CTD is lowered through the water column. If the seabed depth is within the range of the ADCP, the exact distance to the bottom can be measured by the downward-facing ADCP.

3.2.3.3 Small Handheld ADCPs

Smaller, handheld ADCPs contain a compass, battery, and small PC for internal storage of data (Figure 3.37). These ADCPs must be started and stopped manually. The data are downloaded from the internal PC and into the processing software after retrieval of the units. These are typically mounted on underwater rigs, ROVs, or stationary platforms (cf. Figure 3.38).

(a)

(b)

Figure 3.37 Two types of ADCPs typically attached to deployed subsurface moorings.
(a) An Aanderaa SeaGuardII DCP Black. *Source:* Reprinted with permission from Aanderaa, and (b) Nortek AquaDopp Profiler. *Source:* J.T. Hestetun.

Figure 3.38 Illustration of current measurement solutions and mounted ADCPs attached to floating and subsurface buoys and a platform on the seabed. *Source:* Reprinted with permission from Aanderaa.

3.2.3.4 Moorings with ADCPs
Lars Asplin, Institute of Marine Research, Norway

A traditional way of measuring currents in a fjord is to anchor a surface or a sub-surface buoy to a mooring and attach one or more current meters to the rope. For instance, using 2–3 current meters of the acoustic doppler profiler-type with a range of 50–100 m will effectively cover larger portions of the vertical profile in a fjord.

Current observations can be relatively expensive. The potential high spatial variability of the current can also make it difficult to get sufficient information on the actual fjord dynamics without numerous current meter moorings. Thus, a strategy to use information from numerical current models coupled with current observations will be the most cost efficient.

3.2.4 Sediment

Sediment traps are used to measure the amount and nature of sinking particulate organic material (POM). Sediment traps are containers with a series of upward-facing lattices that accumulate particulate organic material raining down from the above water column. Sediment traps are used to estimate the biological production or energy fluxes in the overlying waters.

3.3 Suitability of Equipment in Given Habitat Types

Table 3.2 Overview of sampling gears and equipment and the habitats in which they are used.

Sampling method	Habitat				
	Littoral	Sublittoral	Pelagic	Demersal	Benthic
Sample square / Quadrat	X				
Transects	X	X	X	X	X
ROV		X	X	X	X
Video / Image	X	X	X	X	X
Counts	X				
Demersal trawl		X		X	
Pelagic trawl			X		
Beach seine		X			
Phytoplankton nets		X	X		
Zooplankton nets		X	X		
Gillnets		X	X	X	
Traps		X	X	X	
Fyke nets		X			
Angling		?	X	X	
Longline		?	X	X	
Sled/dredge					X
Grab/corer		X			X
Acoustics			X	X	
Satellite/infrared light			X		
Bottles		X	X		
CTD/Rosette		X	X		
Secchi disk/light sensors		X	X		
Current meters/ADCP		X	X	X	

References

Aglen, A. (1994) Sources of error in acoustic estimation of fish abundance, in *Marine Fish Behaviour in Capture and Abundance Estimation*, (eds. A. Fernø and S. Olsen) John Wiley & Sons, New York, pp. 107–133.

Blondeau-Patissier, D., Gower, J.F.R., Dekker, A.G. *et al.* (2014) A review of ocean color remote sensing methods and statistical techniques for the detection, mapping and analysis of phytoplankton blooms in coastal and open oceans. *Progress in Oceanography*, 123, 123–144. DOI:http://dx.doi.org/10.1016/j.pocean.2013.12.008

Brede, R., Kristensen, F.H., Solli, H. and Ona, E. (1990) Target tracking with a split-beam echo sounder. *Proceedings of the International Symposium on Fisheries Acoustics*, 22–26 June 1987, Seattle, WA (USA), ICES.

Cochran, W.G. (1977) *Sampling Techniques*, 3rd edn, John Wiley & Sons, New York.

Demer, D., Berger, L., Bernasconi, M. *et al.* (2015) Calibration of Acoustic Instruments. *ICES Cooperative Research Report*, 326, 133.

Dierssen, H.M. and Randolph, K. (2013) Remote Sensing of Ocean Color, in *Earth System Monitoring: Selected Entries from the Encyclopedia of Sustainability Science and Technology*, (ed. J. Orcutt), Springer, New York, pp. 439–472.

Dragesund, O. and Olsen, S. (1965) On the possibility of estimating year-class strength by measuring echo-abundance of 0-group fish. *Fiskeridirektoratets skrifter, Serie Havundersøkelser*, 13(8), 48–75.

Ehrenberg, J. (1979) A comparative analysis of *in situ* methods for directly measuring the acoustic target strength of individual fish. *IEEE Journal of Oceanic Engineering*, 4(4), 141–152.

Engås, A. and Løkkeborg, S. (1994) Abundance estimation using bottom gillnet and longline – the role of fish behaviour, in *Marine Fish Behaviour in Capture and Abundance Estimation*. (ed. A. Fernø and S. Olsen) Fishing News Books, Oxford, pp. 134–165.

Engås, A., Skeide, R. and West, C.W. (1997) The 'MultiSampler': a system for remotely opening and closing multiple codends on a sampling trawl. *Fisheries Research*, 29(3), 295–298. DOI:10.1016/S0165-7836(96)00545-0

Foote, K., Knudsen, H., Vestnes, G., MacLennan, D. and Simmonds, E. (1987) Calibration of acoustic instruments for fish density estimation: a practical guide. *ICES Cooperative Research Report*, 144: 1–57, 69.

Gunderson, D.R. (1993) *Surveys of Fisheries Resources*, John Wiley & Sons, New York.

Hamley, J.M. (1975) Review of gillnet selectivity. *Journal of the Fisheries Board of Canada*, 32(11), 1943–1969. DOI:10.1139/f75-233

Horne, J.K. (2000) Acoustic approaches to remote species identification: a review. *Fisheries Oceanography*, 9(4), 356–371. DOI:10.1046/j.1365-2419.2000.00143.x

Horne, J.K., Walline, P.D. and Jech, J.M. (2000) Comparing acoustic model predictions to *in situ* backscatter measurements of fish with dual-chambered swimbladders. *Journal of Fish Biology*, 57(5), 1105–1121. DOI:10.1111/j.1095-8649.2000.tb00474.x

Johnsen, E., Pedersen, and Ona, E. (2009). Size-dependent frequency response of sandeel schools. *ICES Journal of Marine Science*, 66, 1100–1105.

Korneliussen, R.J. and Ona, E. (2003) Synthetic echograms generated from the relative frequency response. *ICES Journal of Marine Science*, 60(3), 636–640. DOI:10.1016/S1054-3139(03)00035-3

Madsen, N., Hansen, K.E., Frandsen, R.P. and Krag, L.A. (2012) Development and test of a remotely operated Minisampler for discrete trawl sampling. *Fisheries Research*, 123–124, 16–20. DOI:10.1016/j.fishres.2011.11.016

Meese, R.J. and Tomich, P.A. (1992) Dots on the rocks: a comparison of percent cover estimation methods. *Journal of Experimental Marine Biology and Ecology*, 165(1), 59–73. DOI:http://dx.doi.org/10.1016/0022-0981(92)90289-M

Ockelmann, K.W. (1964) An improved detritus-sledge for collecting meiobenthos. *Ophelia*, 1(2), 217–222. DOI:10.1080/00785326.1964.10416279

Rosen, S. and Holst, J.C. (2013) DeepVision in-trawl imaging: sampling the water column in four dimensions. *Fisheries Research*, 148, 64–73. DOI:http://dx.doi.org/10.1016/j.fishres.2013.08.002

Rosen, S., Jörgensen, T., Hammersland-White, D. and Holst, J.C. (2013) DeepVision: a stereo camera system provides highly accurate counts and lengths of fish passing inside a trawl. *Canadian Journal of Fisheries and Aquatic Sciences*, 70(10), 1456–1467. DOI:10.1139/cjfas-2013-0124

Ruttner, F. (1963) *Fundamentals of Limnology*, University of Toronto Press, Toronto.

Salvanes, A.G.V. (1991) The selectivity for cod (*Gadus morhua* L.) in two experimental trammel-nets and one gillnet. *Fisheries Research*, 10(3), 265–285. DOI:10.1016/0165-7836(91)90080-Y

Simmonds, J. and MacLennan, D.N. (2005) *Fisheries Acoustics: Theory and Practice*: John Wiley & Sons, New York.

Smith, T.D., Gjøsæter, J., Stenseth, N.C. *et al.* (2002) A century of manipulating recruitment in coastal cod populations: the Flødevigen experience. *Proceedings of the ICES Marine Science Symposia, 2002.* pp. 402–415.

Stenseth, N.C., Bjørnstadf, O.N., Falck, W. *et al.* (1999) Dynamics of coastal cod populations: intra- and intercohort density dependence and stochastic processes. *Proceedings of the Royal Society of London. Series B: Biological Sciences*, 266(1429), 1645. DOI:10.1098/rspb.1999.0827

Tranter, D. and Fraser, J. (1968) *Zooplankton Sampling, Monographs on Oceanographic Methodology 2*, UNESCO, Paris.

Venrick, E. (1978) Sampling techniques, in *Phytoplankton Manual*, (ed. A. Sournia) Page Brothers Ltd/UNESCO, Norwich, pp. 33–67.

Weikert, H. and John, H.C. (1981) Experiences with a modified Bé multiple opening-closing plankton net. *Journal of Plankton Research*, 3(2), 167–176. DOI:10.1093/plankt/3.2.167

Wiebe, P.H., Morton, A.W., Bradley, A.M. *et al.* (1985) New development in the MOCNESS, an apparatus for sampling zooplankton and micronekton. *Marine Biology*, 87(3), 313–323. DOI:10.1007/BF00397811

Williams, K., Towler, R. and Wilson, C. (2010) Cam-trawl: a combination trawl and stereo-camera system. *Sea Technology*, 51(12), 45–50.

Zhao, W.J., Wang, G.Q., Cao, W.X. *et al.* (2014) Assessment of SeaWiFS, MODIS, and MERIS ocean colour products in the South China Sea. *International Journal of Remote Sensing*, 35(11–12), 4252–4274. DOI:10.1080/01431161.2014.916044

4

Sorting Specimens and Preserving Materials

Anne Gro Vea Salvanes, Henrik Glenner*, Jennifer Devine,
Jon Thomassen Hestetun, Mette Hordnes, Knut Helge Jensen,
Frank Midtøy and Kjersti Sjøtun*

4.1 Sampling Diary

During a field study, it is a good habit to write a diary in a book with a hardcover detailing all events that have happened. Alternatively, it is possible to keep a narrative of each day's events in a text document (e.g., MS Word, Libreoffice). This should include information on: what was done, where and when, which gear was used, whether everything worked as it should, if there were problems including what and why, if known. If the station needs to be repeated, one should record which samples were collected, whether any were preserved or frozen, and whether any improvements can be made in the future. Doing this will make it easier to remember key points when analyzing samples and data for making improvements for future sampling events. Depending on the scope of the survey, many types of activities and oceanographic data, for example, may be logged either automatically or by personnel responsible for particular instruments. This sort of data is often vital. For scientific cruises, a summary (not necessarily including all information) of collected data is usually presented in the form of a scientific cruise report.

4.2 Sorting and Preserving Littoral Collections

Given the ease of direct observation in the littoral zone compared to pelagic and benthic habitats, identification in the field, without removal, is possible for many organisms for biodiversity or environmental surveys. In cases where identification *in situ* is difficult, it is possible to bring back samples of flora or fauna specimens for further examination, using a stereomicroscope or other means. If identification is to be done right after the fieldwork, specimens may simply be kept in seawater.

For faunal specimens, fixation can be done using standard media such as in formalin (histology) or ethanol. For identification purposes, seaweed samples

---

* Lead authors; co-authors in alphabetical order.

Marine Ecological Field Methods: A Guide for Marine Biologists and Fisheries Scientists,
First Edition. Edited by Anne Gro Vea Salvanes, Jennifer Devine, Knut Helge Jensen,
Jon Thomassen Hestetun, Kjersti Sjøtun and Henrik Glenner.

may be fixated in formalin, for example, a 4% solution buffered by borax, or dried on a herbarium sheet. If the individuals are sampled for molecular studies, special conservation should be applied. Clean pieces of seaweeds should be dried as soon as possible with silica gel and stored in airtight vials. Animals should be fixated in ethanol or RNAlater*, depending on later use (see Section 4.7). In the case of snails and animals with an exoskeleton, puncturing the shell may be required to allow the fixing agent to penetrate into the specimen for complete fixation. All samples in a collection should be properly labeled and should include information about, for example, collector, locality, time of collection, and other relevant information. For liquid fixation, a standard method is to label both the outside of the container as well as putting a proper quality label inside, with the sample information written by pencil.

Some types of studies involve biochemical analyses of individuals. In such cases, the sample material needs to be transported to the laboratory as fresh as possible in a cool box. The material can then be processed directly or preserved and stored for later analyses. The method of preservation depends on the type and protocol of the biochemical analyses, but often the material can be dried or frozen.

Collecting Samples for Population Biology Analyses

Sampling individuals in the littoral community for demography measurements, size relationships or individual density, requires a destructive method of sampling. Individuals of a population can be sampled in fixed areas defined by sample plots, brought to the lab and processed. Normally, length measurements, weights and age (if possible) are recorded. Populations of intertidal species can be sampled during low water, whereas scuba diving is required for sampling, for example, sublittoral kelp populations.

4.3   Sorting Zooplankton

Zooplankton sampled for examining species composition must be preserved in neutralized 4% formalin solution, while zooplankton sampled for measures of total species biomass can be frozen. If samples are collected for both biomass and species composition, the sample must be divided using a double chambered **plankton splitter,** which will split the sample in two equal halves. A standard procedure for sorting and analyzing zooplankton samples for biomass and species composition developed by the Institute of Marine Research (IMR) in Norway and used internationally since the 1980s (Melle, Ellertsen, and Skoldal, 2004) is described in Box 4.1.

The **volume of seawater filtered by each plankton net** during the sampling will be needed to quantify **biomass**. The volume of seawater is measured by the flowmeter mounted in the front of the net. The biomass can be calculated for each fraction of zooplankton (e.g., size fractions >2 mm, 1–2 mm, <1 mm, krill, shrimps, and fish larvae). Biomass is standardized to **units of weight per volume seawater,** which allows for comparisons of depth distributions of the various fractions of zooplankton or analysis of day and night differences. To standardize the biomass, the weight of each size fraction is divided by the volume of seawater, as estimated from the flowmeter recordings. Units are commonly in gram/cubic meter, (g /m^3) or (gram/10 m^3).

1) All zooplankton in the sample are rinsed into a numbered codend on the zooplankton net and then brought into the laboratory onboard the ship. 2) The sample is emptied into a white sorting tray. 3) Jellyfish are picked out, identified, and their volume is recorded. 4) The remaining sample is next emptied into a plankton splitter and split in two halves; one half is used for biomass determination and the other for species identification and counting. 5) The part of the sample that is used for species identification is preserved in buffered (with borax) 4% formalin until sorting can be

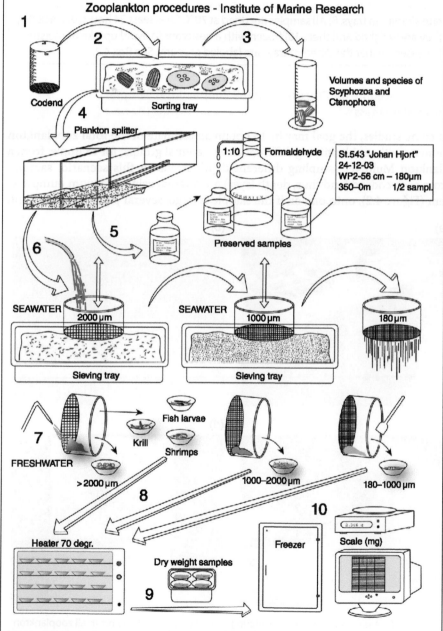

Figure 4.1 Procedures for processing zooplankton samples. See text in Box 4.1. *Source:* Reprinted from Melle, Ellertsen, and Skoldal, (2004) with permission from Fagbokforlaget.

completed; properly labeled (see Box 4.2) plastic bottles with double lids are used. The buffered formalin solution can be made by adding 1 part 40% formalin to 9 parts seawater; for each 100 ml, add 1 ml borax powder and shake well. 6) The sample that will be used to estimate total biomass is sieved through successfully smaller mesh sizes (e.g., 2000 μm, pertains to organisms of sizes >2 mm; 1000 μm, organisms 1–2 mm; and 180 μm, organisms <1 mm in size). 7) The organisms on each sieve are then rinsed with freshwater to reduce the salt content of the adherent water and transferred to small pre-weighed and labeled aluminum trays. From the 2000 μm sieve, all krill, shrimps and fish (mainly fish larvae) are removed and placed on separate aluminum trays. 8) All samples are dried at 70 °C for at least 24 hours. 9) Once dry, they are weighed and then, 10) frozen with a desiccant to avoid uptake of moisture. It is essential that the different trays are labeled properly. See Box 4.2.

4.3.1 Procedure for Processing Small Zooplankton Samples for Total Biomass

In some studies, the goal may be to obtain an estimate of the total zooplankton biomass without identifying different size or animal groups, for example, from a plankton MultiNet sampling different depth strata. In this scenario, samples from each codend (and thus, depth layer) are filtered, using a vacuum pump system (Figure 4.2), onto one or, if the sample is large, several pre-weighed carbon

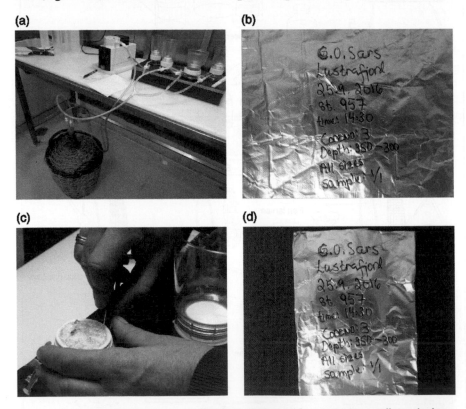

Figure 4.2 (a) A simple vacuum pump filtering system used for processing small zooplankton samples for total biomass estimation. (c) Removing a filtered sample from the vacuum pump. (b and d) Examples of how to label zooplankton samples after filtration. *Source:* A.G.V. Salvanes.

Box 4.2 Labelling zooplankton samples

All samples should be properly labeled so that they can be traced back to a sampling event. The procedure should be as follows:

- Use a permanent marker to write on the aluminum foil that will wrap the frozen samples. Wrap each sample or size fraction separately.
- For samples in formalin, use pencil on paper labels, one of which is placed inside the bottle, the other is affixed to the outside.
- Labels the samples or size fraction with:
  1) Ship name: (for example, R/V "G.O. Sars")
  2) Study site or geographical position: (e.g., Lustrafjord or latitude and longitude)
  3) Date
  4) Time
  5) Station number
  6) Codend number
  7) Depth layer
  8) Size fraction: e.g. >2 mm or 1–2 mm, or <1 mm, fish, krill, or shrimp
  9) Sample: 1/1 if sample was not split, ½ if split in 2, ¼ if the splitter was used twice, and so on.

---

filters (e.g. glass Whatman microfilter). After which, the filters must be wrapped in labeled aluminum foil. See Box 4.2. If multiple filters are used, all filters from one depth-layer sample must be wrapped into the same bit of aluminum foil with the number of filters used noted on the label. After all samples from all depth layers at one station have been processed, place all wraps into one plastic bag, labeled with the station information or number, and freeze at −20°C. Once back on shore, follow the same drying and weighing procedure as described in Box 4.1. Remember to deduct the filter weights from the sample weights when calculating the biomass.

4.4   Sieving and Sorting Benthic Samples

Since benthic sampling methods do not distinguish between organic and inorganic material, a sorting procedure is absolutely necessary. Living organisms collected from hard substrates are often easily sorted out via visual inspection of the sample material. Soft-bottom material, on the other hand, almost always requires several sieving steps to concentrate the biological material, which often is of minute size and requires a dissection microscope to observe. The number of steps to process the sample depends on the animal size fraction of interest. Many environmental surveys focus on two size fractions of animals, for example, organisms larger than 5 mm (megafauna) and organisms larger than 1 mm (macrofauna). In some cases, meiofauna, organisms smaller than 1 mm, is the focus and, as such, a third sieving step must be included.

Figure 4.3 Using seawater to sieve a sample through two different size-fraction sieves stacked on top of each other, where the largest mesh size is placed on top. *Source:* H. Glenner.

A typical soft-bottom sampling situation is when the sampling gear has returned a rather muddy sample that has to be sieved (Figure 4.3). Sieving using seawater is required to wash the sediment from the sample. General care needs to be taken to soften the water flow to avoid damaging organisms more than necessary. The water should not deviate too much in temperature from that of the collection depth, especially in warmer climates, because organisms may dissolve. For tropical sampling, this might entail pumping water from greater depth rather than using warmer surface water. Needless to say, freshwater should never be used to wash marine organisms.

Several auxiliary methods can be employed to extract biota from sediments. As many organisms are lighter than the sediment, it is often useful to float off and sieve the top fraction first; removing as many organisms as possible early in the sieving process minimizes specimen damage. Unsieved samples are sometimes stored in a bucket or trough overnight in a cold room. As oxygen concentration in the sediment decreases due to respiration, animals will exit the sediment and often climb up the sides of the container, allowing easy collection of undamaged specimens.

4.5 Fish and Nekton

4.5.1 Trawl Samples

There are several steps that must be taken to ensure that the catch from each net and each station does not mix. On a research ship, the crew will put the catch on deck, but the scientists and students will be responsible for collecting and

Figure 4.4 Sampling with a MultiSampler. (a) Preparing to set a pelagic trawl fitted with a MultiSampler and 3 codends. (b) The MultiSampler trawl hauled on deck. (c) A typical mesopelagic catch in Masfjord, western Norway. (d) A night catch from three codends; the catch in T1 from the deepest mesopelagic layer, in T2 between two mesopelagic layers, and in T3 from the shallowest mesopelagic layer. *Source:* A.G.V. Salvanes (a, b, and c) and F. Midtøy (d).

processing the samples. If a MultiSampler is used to sample populations from different depths of the mesopelagic community, more than one codend and one sample will be on deck at the same time (Figure 4.4). *Samples from each codend must be properly labeled* and put into *separate* containers to avoid cross-contamination. It is useful to have pre-made labels encased in hard plastic envelopes (e.g. with labels T1 to T3 if 3 trawl codends are on the MultiSampler). If a trawl sample is from a bottom tow, it is intuitive to use labels BT. Typically, bottom trawls have just one codend (Figure 4.5, Figure 4.6).

For each codend there will be two types of data:

1) Station data: includes, for example, ship, geographical position (longitude and latitude), station number, date, depth, gear type, start tow time, haul tow time, nautical miles towed, tow speed, ship log. The time unit used on a research ship may differ depending on the standards of that nation; however, one commonly used time is Greenwich Mean Time (GMT) or, as it is otherwise known, Coordinated Universal Time (UTC). Expressing time in local time units is

Figure 4.5 A typical catch from a bottom trawl from 490 m depth in Masfjord, western Norway. (a) The trawl after being hauled onboard. (b) The trawl catch on deck, prior to sorting into baskets of single species. (c) Weighing the baskets of individual species in the wet lab and recording the data. (d) A roundnose grenadier (*Coryphaenoides rupestris*) on a measuring board. Grenadier tails are fragile and easily break off in the trail, therefore the standard length measurement is from the tip of the snout to the posterior edge of the dorsal fin. *Source:* A.G.V. Salvanes.

useful, particularly if diurnal changes in light influence the distribution of the studied organisms. During European summer time, local Norwegian time is GMT + 2 hours, whereas during the winter the local Norwegian time will be GMT + 1 hour. On a large research vessel, the station data are recorded automatically on standardized forms or data sheets received from the wheelhouse (bridge) or using survey logging programs. Different countries have different design of these forms or systems for recording ship data, but the information recorded will generally be the same.

2) Biological data. It is useful to bring forms, which are formatted for your purpose of sampling. Waterproof paper will be useful for outdoor work, and pencils must be used for writing (text written by fiber tip or ballpoint pens on wet paper will soon disappear!). Electronic measuring boards and systems for registering the catch are also widely used. This allows downloading the data directly (or through database queries) into the format needed for analysis.

Figure 4.6 A typical catch from a bottom trawl from 90–150 m depth in the North Sea. (a) The catch is hauled on deck and placed into a large container before entering the fish lab. (b) The catch is placed into baskets (still unsorted at this stage) to be weighed; (c) the catch is sorted and measured; (d) a basket containing only saithe (*Pollachius virens*). (e) A hake (*Merluccius merluccius*) on an electronic measuring board. *Source:* H. Saivolainen, University of Bergen, Norway and F. Midtøy.

4.5.1.1 Sorting a Codend Sample and Subsampling

Subsampling of the catch is used when large samples, too large and too time consuming to process fully, are taken from marine populations. Care should always be taken to never sample more than is needed because of the harm done to organisms and/or the habitat, but sometimes, large samples are unavoidable. Many research surveys must prove that damage to organisms (mortality) and habitat will be kept to a minimum before being granted permission to survey (e.g., Netherlands must apply for permission to survey and provide estimates of expected mortality (number of organisms killed); once this estimate is reached, surveying must stop).

Subsamples are when only part of the catch is fully sorted and sampled for the study. The principles of subsampling using a midwater catch targeting mesopelagic organisms in the DVM is illustrated in Figure 4.7 and Box 4.3, while Box 4.4 details how to estimate total catch from the subsampled catch. On modern fisheries research surveys, some of this work is automated by electronic recording and database programs.

Box 4.3 How to subsample the catch and take measurements

Using a sample from one trawl codend targeting the mesopelagic community in a Norwegian fjord, the Masfjord, we outline the general principles for processing the catch (Figure 4.7). The general principles will be the same for catches from bottom or pelagic trawling.

1) Containers for holding the catch (e.g., baskets, buckets) should be labeled with the gear used; for bottom trawl catches, the label may be BT, while for MultiSampler codends, labels will be based on the sample depth (e.g., T1 for deep, T2 for mid, and T3 for shallowest). If the catch is large, more than one

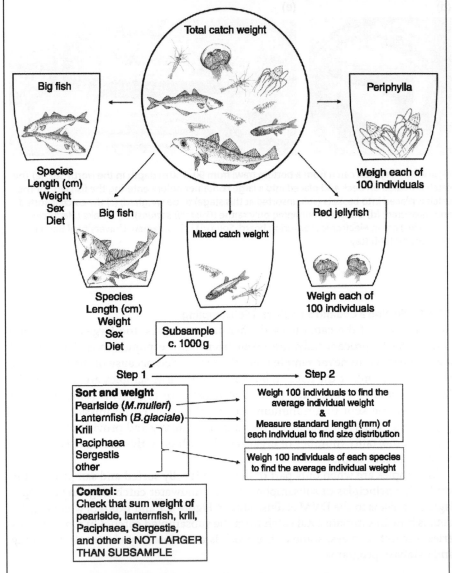

Figure 4.7 How to subsample and measure the catch of a mesopelagic community.
Source: Artwork by R. Jakobsen.

container may be needed for the catch from each codend. Hard plastic (pre-made) labels are often useful for this.

2) Always take steps to avoid mixing the catch from multiple tows, gears, or codends at all steps in the procedure.

3) Sort large fish into separate containers, using a separate container for each species.

4) Sort large medusae, such as the species *Periphylla periphylla* and red jellies (*Cyanea capillata*) into separate containers.

5) Weigh the contents of each container and write this information on the catch recording form (or enter it into the electronic measuring board). The sum of all containers will be the *total catch weight*.

6) Weight of the mixed catch (here, the mesopelagics) will be the total catch weight minus the weight of the large fish and jellyfish (i.e., all species sorted into individual containers). If the mixed catch weighs more than one kilo, take a **random subsample** of about 1000 grams from the mix of organisms. *If there are multiple containers per codend, ensure that the catch is well mixed. Containers should not contain only krill or only fish, but be representative of the entire catch.*

> *What is a random subsample?*
> *If a sample is not a random subsample, can*
> *we use it for the study?*

7) Weigh the subsample. Write this weight onto the catch recording form (Figure 4.11). Identify this as **subsample weight**. Failure to do so will result in errors when calculating catch numbers and weights later.

8) Sort the subsample into separate species (Figure 4.8). The mesopelagic assemblage may include the following species: pearlside, *Maurolicus muelleri*, lanternfish, *Benthosema glaciale*, krill, *Meganyctiphanes norvegica*, and the pelagic shrimps, *Paciphaea* spp. and *Sergestes* spp. Most other species will be in smaller numbers and may be difficult to identify to species. If identification is not necessary, this could be grouped into one group, classified as 'other'. Note: using an 'other' group may limit the type of detailed analysis one can conduct with the data later on, so be judicious in the use of this category.

9) Weigh each species of the subsample.

10) Control: calculate the sum of the weights of all species in the subsample as a control of correct weighing.

11) Measure the length (e.g., standard length, from snout to the end of the caudal peduncle/beginning of the caudal fin) of 100 of each of the fish species from the random subsample. The measurement unit here is mm: round down to the nearest mm (i.e., 30.9 mm = 30 mm). If less than 100 individuals of a species are in the subsample, measure all specimens. Write this information on a length-measurement form (Figure 4.13). If using paper forms, the form should include station number, codend number, species, measurement unit (here, mm), and name of the person who does the measurements; electronic measuring boards will automatically link the individual data to the station data, but

Figure 4.8 Sorting the mesopelagic catch in Masfjord. (a) A typical catch from a 5–10 min haul. (b) The most often caught species: pearlside (*Maurolicus muelleri*), lanternfish (*Benthosema glaciale*), *Sergestes* spp., *Pasiphaea* spp., krill (*Meganyctiphanes norvegica*), and the helmet jellyfish (*Periphylla periphylla*). (c) Sorting a subsample. Counting of (d) pearlside and (e) *Sergestes* spp. by placing in piles of 10 individuals. (f) Measuring individuals and recording standard length (SL). *Source:* A.G.V. Salvanes.

the user will need to key in species and ensure the measurements are recorded in the correct units (e.g., cm, half cm, mm). Record the weight of the individuals measured for length (Figure 4.13).

12) For small invertebrates, count out up to 100 of each species (e.g., *M. norvegica*, *Pasiphaea* spp., *Sergestes* spp.) and record the weight for each of the species. Average individual weight can then be estimated.

13) For the Masfjord data, weigh individual *Periphylla periphylla* for up to 100 individuals. If less than 100 were caught, all should be weighed. Record the weight of the measured individuals. Note: total weight of all *P. periphylla* caught should have been already recorded (step 5).

14) Count 100 other large jellyfish species, such as the red jellies (*Cyanea capillata*) and record the weight of those counted by species.

15) For the large fish species separated into separate containers in step 3: measure the total length (for most species; from snout to the end of tail held in natural position; see Figure 4.9 for exceptions) of up to 100 individuals of each species; here, the measurement unit is rounded down to the nearest cm (e.g., 30.9 cm = 30 cm). For the Masfjord sample, the procedure is to convert this to mm when entered into the data file for individual length measurements

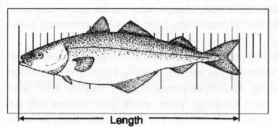

Usually, the length of fish is measured as total length, i.e. from snout to the end of the tail.

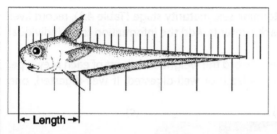

Grenadiers (Isgalt/Skolest) are measured from the snout to the anterior edge of the anal fin.

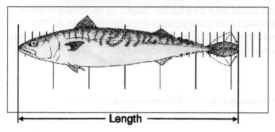

Rabbitfish (Havmus) are measured from the snout to the posterior edge of the dorsal fin.

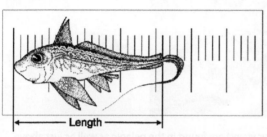

Mackerels and horse mackerels are measured from the snout to the end of the tail with lobes squeezed together.

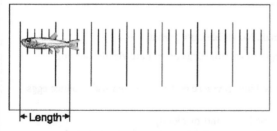

Mesopelagic fish are measured from the snout to the boneknob at the tailroot.

Figure 4.9 How to measure the length of various fish species at the Institute of Marine Research, Norway. *Source:* Artwork by R. Jakobsen.

of all species. To reduce potential errors, always note the measurement unit on the paper form. For electronic measuring boards, this is recorded automatically in the database.

16) Full biological sampling may be wanted for a subset of the species. Taking the Masfjord example, this information is wanted for the large fish predators: *Pollachius pollachius* (pollack), *Pollachius virens* (saithe) (Figure 4.10) and *Micromesistius poutassou* (blue whiting). Here, extra information will be collected from 30 individuals of each species. For these individuals, in addition to length, record individual weight, take out the otoliths, and then open the belly of the fish to determine sex, maturity stage (Table 4.1), record liver weight, and identify stomach contents. Add the information to an individual form (Figure 4.14)

a) Diet: Identify prey species. Record the weight and count of each prey type. Record whether the prey is fresh or well-digested. If well-digested, one

(a)   (b)   (c)

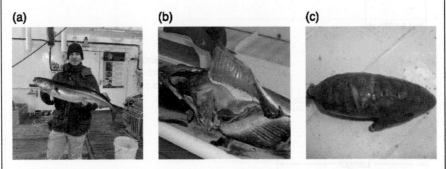

Figure 4.10 Large saithe (*Pollachius virens*) are found in the pelagic as well as just above the seabed in deep west Norwegian fjords. (a) A 5 kg saithe caught by hand-line at 90 m depth, immediately below the shallowest mesopelagic sound scattering layer. (b) The individual with its belly opened and the stomach removed. (c) The stomach of the fish was packed with pearlside (*Maurolicus muelleri*). *Source:* A.G.V. Salvanes.

Table 4.1 General description of maturity stages of fish (Mjanger *et al.* 2016).

Stage	Description
Blank	Undecided/not checked
1	Immature. Gonads are small. No visible eggs or milt.
2	Maturing. Gonads are larger in volume. Eggs or milt are visible, but not running.
3	Spawning. Running gonads. Light pressure on the abdomen will produce eggs or milt.
4	Spent/Resting. Gonads small, loose, and/or bloody. Regeneration starting, gonads somewhat larger and fuller than stage 1. No visible eggs or milt.
5	Uncertain. Only use when difficult to distinguish from stages 1 and 4.

Table 4.2   Condition of gall bladder and hind guts used to differentiate between empty and regurgitated stomachs of fish.

Stage	Gall bladder	Bile color	Hind gut	State
1	Shrunken, empty or with small amount of bile	Pale	Contains large amounts of bile and digested food material	Feeding[a]
2	Elongate	Pale green to light emerald green	Contains some bile and digested food particles	Feeding[a]
3	Elongate	Dark green	Empty or contains some food particles	Empty
4	Round	Dark blue	Empty	Empty

[a] If fish satisfying these criteria are found without food in their stomach they should be classified as regurgitated.

should be able to approximately count the number of unidentified fish by counting backbones and/or otoliths (2 per fish), which can be preserved for later identification to species. If stomach contents are fresh, identify prey to species and, for mesopelagic fish prey (e.g., pearlside and lanternfish), measure standard length in mm. Many predators could have regurgitated their prey. This occurs because the gas in the swimbladder will expand when fish are brought to the surface from greater depths, forcing the stomach contents (and possibly the internal organs) out through the mouth. Even when fish have regurgitated, one can examine the gallbladder and bile color (Table 4.2) to check whether the fish had recently eaten or had an empty stomach.

17) Make thorough records of the data at every step. A useful practice is to designate clear shift roles, that is, each person is responsible for a particular activity, such as the responsibility for collating all data sheets and ensuring the data records are complete. For paper records, the data need to be entered into for example, spreadsheets for later analysis.

18) For paper records, write the total weight of the catch, the total weight of the mix (unsorted part of the catch), and the weight of the subsample taken from this mix on the station form (Figure 4.11). Also note the weight and number of each species in the subsample. These notes will be important when using the subsample data to calculate the total catch of each species from each trawl haul codend. For electronic measuring boards, most of this information has already been logged in the measuring boards. The only missing information may be the weight of the unsorted mixture and the weight of the subsample; this information must not be misplaced since it is needed to calculate the total weight and number for the species in the subsampled catch.

Box 4.4 Converting subsampled catches to total catch

If the subsample was truly random (came from a well-mixed catch and therefore contains all species and size ranges as in the total catch), then the subsample will consist of the same proportion of each species (in weight and numbers) as in the unsorted mixed catch.

In the Masfjord example, the mixed catch was made up of mesopelagic organisms after all large fish and jellyfish species were removed. In this example, we converted the subsampled catch to total catch for both weights and numbers.

An example

After removing the large fish and jellyfish species, the remaining unsorted (mixed) catch weighed 5000 grams. A subsample weighing 975 g was taken for further sorting. Species in the subsample were pearlside, *Maurolicus muelleri*, lanternfish, *B. glaciale*, krill, *Meganyctiphanes norvegica*, and the pelagic shrimps, *Sergestes* spp. and *Pasiphaea* spp. The weights of each of these species (species groups for shrimps) in the subsample were, respectively, 117 g, 398 g, 316 g, 98 g, and 46 g.

1) To find the proportion by weight of each species in the subsample:

$$Proportion\ of\ species\ A = \frac{weight\ (g)\ of\ species\ A\ in\ the\ subsample}{total\ weight\ (g)\ of\ the\ subsample\ (all\ species)},\ \text{where}$$

species A refers to one species in the subsample.

The proportion of pearlside in the subsample was 117/975=0.12 (i.e., the subsample was 12% pearlside by weight).
The proportion of lanternfish in the subsample was 398/975=0.41 (i.e., the subsample was 41% lanternfish by weight).

2) To estimate the total weight of each species in the unsorted, mixed part of the catch:

$$Total\ catch\ of\ a\ species = Proportion\ of\ a\ species \times Weight\ of\ the\ mixed\ catch$$

The total catch of pearlside was 0.12×5000=600 grams. This number (total catch of pearlside) can now be entered into the station form (Figure 4.11).

Estimate the total number of each species in the mixed catch

1) Using the weights recorded from the fish measured for length in the sample (Figures 4.7 and 4.13), estimate the average weight of an individual for each species in the subsample.

$$Average\ weight\ of\ species\ A = \frac{weight\ (g)\ of\ species\ A\ that\ were\ measured\ for\ length}{number\ of\ species\ A\ that\ were\ measured},\ \text{where}$$

species A refers to one species in the subsample.

For *B. glaciale*, 100 individuals were measured (see Figure 4.13) and these 100 fish weighed 220 g. The average weight of an individual was 220/100 = 2.2 grams.

2) Calculate the total number for each of the mesopelagic organisms in the catch (i.e., in the unsorted mix).

$$Total\ number\ of\ species\ A = \frac{total\ catch\ (g)\ of\ species\ A}{average\ individual\ weight\ of\ species\ A}$$

The total catch of *B. glaciale* in the unsorted mix was 0.41 × 5000 = 2050 g. The total number of lanternfish was then estimated to be 932 individuals.

4.5.2 Sorting Hook-and-Line Samples

When a fish is hauled on the deck, quickly kill the fish by striking the head with a blunt (moderately heavy) instrument; this is because care must be taken to minimize fish suffering. Label the fish with an individually numbered tag or water-resistant paper. Labels can be affixed to the fish by stapling the label to the gill cover or by placing the fish and label in an individual container (1 fish per container) or, if freezing the fish, in a bag (see Box 4.5, procedures for freezing fish). At a minimum, the following information should be noted for each fish:

- **Unique identity code**: for example, name of person who captured the fish, fish number, station number.
- **Capture depth**
- **Time of capture**

If the fish is to be kept alive after capture, fish can be tagged with alternative tags, such as T-bar, spaghetti, or PIT tags, after anaesthetizing the fish. In most countries, one must have the proper certificates (animal welfare, permission) beforehand to house animals alive to perform tagging and to conduct experiments on live organisms.

After fishing is finished, the catch must be processed the same way as for large fish in Box 4.3.

4.6 Data Records

For students and early career scientists who conduct field studies to answer questions and test hypotheses, whether it be for their own research or to learn a technique, learning the basic procedures of recording data is essential. Understanding and learning the basic procedures will make it easier to determine how to analyze and interpret the data. One should always know how the data for their research is collected to fully understand the assumptions and potential biases that may be inherent in the process. Ideally, one should take part in the data collection at least once.

Automatic electronic measuring is becoming increasingly common at large marine research institutes, where one of the main activities of the institute is fish

---

Box 4.5 How to freeze samples

Write all required information (see Box 4.3 and Figure 4.12) on a piece of paper using a pencil. The paper should be of a type that does not disintegrate when wet. Do not use a ballpoint pen or marker because the ink will run when wet, making the label unreadable. Place the paper label in a sealed plastic bag and place this inside the bag with the sample. Be sure that the information is visible through the plastic bag before you freeze the sample (Figure 4.12). A copy of the station form (Figure 4.11) with the species name and/or number of organisms can be used. Fish should ideally be frozen flat. If individuals are small, freeze them in small amounts or, if large, freeze them individually so that small amounts can easily be taken out of the freezer. Small amounts or individual fish are easier to thaw when the organisms are worked with later. For freezing zooplankton, see Section 4.3.

(a)

Required information for trawl sample label:

Ship name:
Site name or geographical position:
Station/Serial number:
Species:
No organisms:
Date:
Time:
Your name:

(b)

Figure 4.12 (a) Example of label information when freezing a sample. (b) Frozen samples should be flat packed and the label must be visible through the plastic bag. *Source:* A.G.V. Salvanes.

---

stock monitoring. Here, we detail working with paper records, using procedures used in the field course in Masfjord and Lustrafjord run by the Department of Biology, University of Bergen in Norway, as an example.

4.6.1 Station Records and Species Composition

A typical form used for processing the catch should contain fields for recording station catch, and information about the species composition of the catch in weight and numbers (Figure 4.11). Before sampling, if on large research vessels, the form can be given to the bridge, who will fill in some of the station data. Alternatively, one may have to stand on the bridge and record this information; this depends on the normal procedures of the vessel crew. One form should be used for each station and gear; if the gear has multiple codends, a separate form should be used for each. The field course uses a unique station number and serial number (a unique number that provides information about region and haul) for each codend. Procedures for assigning unique station codes may differ for other surveys and/or institutes. The upper part of the form contains the station information, which are data that identify and characterize the sampling site,

HAVFORSKNINGSINSTITUTTETS KVALITETSSYSTEM
Senter for marine ressurser, Håndbok for prøventaking av fisk og krepsdyr   FISKESTASJONSSKJEMA (S)

ÅR	LAND	SKIPSKODE	SKIP	MND	DAG	ST.NR.	SERIE.NR.	ST:TYPE	POSISJON BREDDE	LENGDE	NSøV	SYSTEM	OMR.	LOK.	BUNNDYP
13	58	4	GS	5	10	109	23115		67°40.1'	012°56.2'	1	2	56	19	

| REDSKAP | | | RETNING | FART (KNOP) | START | | STOPP TID (UTC) | DISTANSE | TILSTAND | KVALITET | FISKEDYP | |
ANT	KODE	NR.			TID	LOGG					MAKS	MIN
1	3513	1	29	3	1308	9047	1338	15	1	1	420	400

Skip, hvor prøven er tatt:
Fartøy: ....
Reg.nr: ....
Skjema, ført og kontrollert av:
Prøvetaker: ....
Punchet: ....

STATION FORM BIO325			Station/Ser:		Date:	
TOTAL CATCH WEIGHT:						
MIXED CATCH WEIGHT:			Site:			
SUBSAMPLE WEIGHT:						

| | TOTAL CATCH | | SUBSAMPLE | | Your name: Length sample | |
SPECIES	WEIGHT (G)	No.	WEIGHT (G)	No.	WEIGHT (G)	No.

Figure 4.11 An example of a station form used during the field course in Masfjord, Norway. The top part of the form contains station information, while catch information is recorded at the bottom.

often logged by instruments. In general, every time gear is deployed, a new station and serial number will be created. Below the station information, the form contains fields for recording biological data. One line is used for each species; individual data are not recorded on this form, but length measures are noted on a form such as the one in Figure 4.13 and individual data for large fish are noted on a form like the one shown in Figure 4.14.

Ship: H. Mosby	Station/Ser no: 109 / 22140	Date/Year: 17.9.2015
Species name: B. Glaciale		Your name: C. Berntsen

		count						notes
0					0			
1					1			
2					2			
3					3			
4					4			
5					5			
6					6			
7					7			
8					8			
9					9			
3 0					0			
1					1			
2					2			
3					3			
4					4			
5					3			5			
6					6			
7					7		100 individuals
8			1			8		weigh 220 g
9			1			9		
4 0			1			0		
1					1		UNIT measured: mm
2					2		
3					3		
4					4		
5						4			5		
6				2			6		
7					7		
8	HHT HHT			12			8		
9			1			9		
5 0	HHT					9			0		
1					3			1		
2	HHT				8			2		
3	HHT		6			3		
4	HHT				8			4		
5	HHT HHT					14			5		
6				2			6		
7	HHT		6			7		
8	HHT			7			8		
9				2			9		
6 0	HHT		6			0		
1					1		
2				2			2		
3					3		
4					4		
5					5		
6					6		
7					7		
8				2			8		
9					9		
	SUM	100						

Figure 4.13 An example of a length measurement form used during the field course in Masfjord, Norway. Measurement units can be in either mm or cm. One form is used for each species. Data should also include the number and total weight of the measured fish.

Individual form for large fish			Date:				**If prey is easy to measure, DO IT!!, but remember to write on length**
Ship:							form that this is prey to which fish ID
Species:				Station/series no:			Date/Year: Your name:
						yes/no	
Fish no.	Length (cm)	Weight (g)	Liver weight	Sex (m or f)	Fishing depth	Digested??	Diet: type and numbers + total prey weight for each prey species
							Tot numbers Tot weight
							M.maurolicus:
							B.glaciale:
							M.norvegica:
							Sergestes spp.:
							Pasiphaea spp.:
							M.maurolicus:
							B.glaciale:
							M.norvegica:
							Sergestes spp.:
							Pasiphaea spp.:
							M.maurolicus:
							B.glaciale:
							M.norvegica:
							Sergestes spp.:
							Pasiphaea spp.:
							M.maurolicus:
							B.glaciale:
							M.norvegica:
							Sergestes spp.:
							Pasiphaea spp.:
							M.maurolicus:
							B.glaciale:
							M.norvegica:
							Sergestes spp.:
							Pasiphaea spp.:

Figure 4.14 An example of a form to use for individual biological data for large fish that feed on mesopelagic organisms during the field course in Masfjord, Norway.

The total catch of each species in weight and number is recorded on the station form (Figure 4.11). All data on catch are entered there. The unit used for weight on the station form should be the same for all stations to avoid errors. When working with small individuals, as in the Masfjord example, grams is the recommended unit for weight (1 g = 0.001 kg).

4.6.2 Detailed Individual Measurements

Detailed measurements of individual fish can be taken. Each additional form that is used should include the station information (or a clear way to link back to the station and catch information) and the species name. For field studies in the Masfjord, the students should use Latin genus and species names. Other surveys may use FAO species codes. Regardless of the method used, the name should not be confusable with any other species.

Length measures can be recorded on **length forms** (Figure 4.13). The measurement unit (e.g., mm or cm) must be noted on the form. Individuals can be frozen or preserved using chemicals for further detailed studies on shore. See Section 4.7 for fixation methods.

Other biological data may be taken on individual fish onboard; this is usually done with larger species or for sizes that do not require microscopes to obtain the information. For these data, information that pertains to one individual is written on one line (see Figure 4.14 for an example of a form to use for individual biological data for large fish that feed on mesopelagic organisms). Data may include a fish ID number, length, individual weight, sex, maturity stage, gonad weight, liver weight, and diet information (i.e., number and weight of each species recorded). Samples for other purposes may also be taken from an individual for example, fin/gill clips for genetics, otoliths for age determination, gonad samples for histology and/or fecundity measurements, or tissues for toxicology. Otoliths are normally placed into envelopes or vials, with samples for genetic analysis added to vials containing RNAlater or ethanol, depending on the analysis method. All samples should be clearly marked with the same fish ID number and station data (e.g., unique station and/or serial number). Without the unique fish ID number and station/serial number, linking individual data in later analyses will be next to impossible. This will also avoid confusing samples from multiple stations. Procedures for sampling and marking the samples should be established and well-defined before fieldwork begins.

4.6.3 Information Transfer to Data Files

Before data collections can be analyzed and interpreted, it is essential to transfer them to a database: the database may be complex or as simple as an Excel spreadsheet or workbook. Information transfer to safely stored electronic databases or files can be done in two main ways: downloaded from electronic measuring boards and scales or manually, where data written onto paper records are transferred manually into a database or spreadsheet.

Here, we will discuss setting up a data frame to house the data, much like one would in Excel or other spreadsheet program (where data have columns and rows). **Each column represents one variable** and **each row represents a single observation**, such as a single fish. Columns should be named so that it is easy to understand the kind of measurement each variable represents.

One recommendation for data format is to record station data and total catch weights and numbers for each species on one spreadsheet, where each row represents a station (i.e., sampling site) and all station and catch information are in columns. Station data typically includes information about time of sampling, geographical position, sampling depth, gear, gear parameters (if measured; e.g., net opening, doorspread, wingspread, towing speed), and/or weather information. Combining station data with catch information in this format allows for a quick overview of species composition, without requiring any special skills (e.g., R programming). For gears that have multiple codends, use one row in the dataset to represent one trawl codend sample. Hence, for a MultiSampler with three codends, there will be three lines on the spreadsheet, one for each codend catch.

Use another spreadsheet for recording the individual measurements of length. One row will include the species name (for the field course in Masfjord, this will be Latin name in one column) and the length of one individual of that species in the next column. The same must be done for individual biological data (e.g., length, weight, sex, maturity status, and eventual diet) on a third spreadsheet. It is wise to include station data (including station number, depth, date, time of day, and gear) in columns at the beginning of the length spreadsheet and on the individual spreadsheet. These represent the link between the station data and the length (or biological) data; the link between station and individual must remain clear. These variables should be given the same name in both files and appear in the same order.

For the case study of mesopelagics in Masfjord, the following variables were chosen for both the **station** and **individual biological** spreadsheets (or separate files): area, year, month, day, station, serial number (unique code for a region and station), latitude, longitude, sampling type (e.g. follow a fixed depth layer, sample a specific depth range), trawl type, start depth, end depth, day or night (because diel vertical migration was studied), local start time, local end time, and which echo-layer was sampled. Geographical position (latitude and longitude) must be in decimal degrees; this is because later analyses require the data in this format to avoid programming errors. The names of these variables must not contain empty spaces, but instead be written as one word; this will be useful later when the data are input into a data analysis program. For example, **start depth** is renamed as *start.depth* and the variable name coding whether it is day or night is named *day.or.night*. The values of this variable were set only to be day, night, or NA (not applicable). NA is used if the value is missing. While it might be intuitive to just leave the field blank, this can sometimes create more work later when reading the data into an analysis package, such as R. Demonstration scripts and data sets are available online at: filer.uib. no/mnfa/mefm/

In our case study, the station file contain data collected from a MultiSampler with 3 trawl codends, where sampling was from either following echo-layers or following a fixed depth range (see Section 2.5.1.5). This variable was called *sampling.type*. Having a separate code for the different types of sampling procedures allows the data to be merged into one spreadsheet for later data analysis. As mentioned in Chapter 2, one must always **avoid pooling or combining data from multiple sampling procedures** because different ways of sampling will be designed to answer different questions.

Examples of how to organize spreadsheets for large data sets are provided on our web page at the University of Bergen: filer.uib.no/mnfa/mefm/

Once the data have been typed into a spreadsheet or data program, make sure to save a backup of the data. Common practice on large surveys is to have the data stored locally on a server on the vessel, which is then pushed to a database at the institute. Researchers may also make a local backup on an external hard drive. Remember – backup hard drives are cheap; having to replicate a study due to loss of data is not.

Details on how to analyze the data, using data from the field study in Masfjord, is discussed in Chapter 5.

4.7   Samples for Storage

Sample preservation of biological material is crucial to most marine biological work. Once the decision has been made to collect samples for post-processing (i.e., it entails the collection of dead organisms), preservation is necessary. The preservation method used depends on the purpose of the sampling; using the wrong preservation method could ruin the goal of the research. Information about sampling purpose is, therefore, a prerequisite for preservation and post-processing of the collected biological material. Some examples of research aims and their preservation method include:

- Examination of the gross morphology or a specific structure from a certain species. The method of preservation may be freezing, if morphological structures are observable with the naked eye or dissecting microscope, or in ethanol or formalin, if the structures are very small or the specimen will be retained for long periods.
- Immunohistochemistry for evolutionary development purposes may require the fixation with for example, paraformaldehyde.
- Alpha-taxonomy, that is, taxonomic description of a species, ethanol, or formaldehyde might be used.
- Description of the ultrastructural (cellular and subcellular) level will require primary fixation in glutaraldehyde, often followed by secondary fixation in Osmium tetroxide.
- Research into life history (population genetics) will require either freezing sample tissue or preserving it in ethanol.
- In order to use the sampled biological material for genomic and/or transcriptomic studies, flash freezing or RNAlater might be applied.

- Examining intraspecific relationships (predator–prey relationships) may require formalin fixation or freezing the stomach contents for later dissection.
- Community investigations, such as structure and interspecific relationships in a particular habitat, may require large-scale sampling of the habitat where not all the species can be worked up in the field (e.g., sampling benthic invertebrates from the sediments, littoral zone studies). Samples from such studies are typically bulk fixed in buffered formalin or ethanol.

Regardless of preservation methods, samples must be clearly labeled, with all information needed to properly link it to biological and station data. Box 4.5 and Figure 4.8 detail procedures on how to properly freeze fish samples.

Although specific fixation procedures exist for most preservation purposes and one should investigate thoroughly the current methods used for their research questions, some general categories can be provided. Seven major categories of fixatives for biological material, commonly used in marine studies, are described below.

4.7.1 Fixatives

1) **Aldehydes (formaldehyde and glutaraldehyde)** cross-link (bonds that link polymer chains) readily with various biological protein functional groups and act to stabilize the form and structure of the tissue at the cellular level. The number of cross-links increase with time, entailing that tissues will stiffen more with increased contact. If aldehyde is stored for longer periods it oxidizes and forms formic acid. To avoid this, and to generally stabilize the pH during the fixation process, buffer should be added to the solution. Depending on the purpose for fixation and tissue type, different buffers are used, for example, sodium borate, phosphate, calcium carbonate, and even seawater. For electron microscopy based on glutaraldehyde as a fixative, sodium cacodylate is often used as buffer.

   a) **Formaldehyde** is the most widely used fixative for various purposes. This fixative causes moderate tissue shrinkage of cellular structure. Various concentrations are used. A common concentration is 3.7–4% formaldehyde in a water solution with a phosphate buffer and 1% methanol added to prevent spontaneous condensation reactions. The time for formaldehyde to penetrate an average type of tissue is relatively fast (0.5 cm per hour) due to the small size of the molecule. Because formaldehyde is inexpensive and manufactured in large quantities, the fixative is well suited for identification work and standard histology (Thavarajah *et al.*, 2012). Soft-bodied organisms, such as fish, may shrink; however, the shrinking effects can be estimated and compensated for in certain studies (Kristoffersen and Salvanes, 1998).

   b) **Paraformaldehyde for immunohistochemistry.** Commercial grade formalin (37% formaldehyde in water) may often contain methanol that, even when diluted, may inhibit antibody binding due to conformational changes of potential antigen binding sites of proteins. Because of this, preservation solutions prepared for antibody immunohistochemistry should be prepared from paraformaldehyde. Paraformaldehyde is a pure polymerized (solid) form of formaldehyde. Dissolved in water, it becomes formaldehyde in its purest form without additives. To stabilize the fixation process and prevent polymerization (to paraformaldehyde), a phosphate buffer should be added. A frequently used fixative for immunohistochemistry is 4% formaldehyde prepared from solid paraformaldehyde in a phosphate buffer (Kiernan, 2000).

c) **Glutaraldehyde**. Formaldehyde is inferior to the larger aldehyde, glutaraldehyde for ultrastructural studies. However, glutaraldehyde penetrates average tissues at significantly slower rates due to the larger molecular size. Glutaraldehyde possesses two aldehyde groups compared to one in formaldehyde and has, therefore, a larger capability for cross-linking proteins, forming a meshwork that stabilizes the protein mass and preserves the cellular morphology. Glutaraldehyde is therefore superior to formaldehyde in electron microscopy based ultrastructural work. A typical glutaraldehyde solution for use of transmission electron microscopy is 2.5% high quality glutaraldehyde in a sodium phosphate or a sodium cacodylate buffer (Kiernan, 2000).

2) **Alcohol (ethanol)**: denatures protein and is used as a tissue stabilizing agent. Ethanol is readily accessible, inexpensive, and relatively harmless, which makes it useful as a standard fixative and for long-term storage. Severe tissue shrinkage will result because of the dehydration effects of alcohol. Therefore, this type of fixative is not well suited for ultrastructural work or the fixation of soft-bodied organisms. As a compromise between the tissue stabilizing effects and the unwanted shrinkage, 70% ethanol is typically used. Ethanol is used specifically for DNA preservation because the dehydration reduces hydrolytic DNA damage caused by free water. For DNA preservation, high concentrations (96–100%) are used to maximize the dehydration effects.

3) **Flash freezing (liquid nitrogen)**: ensures that tissue samples will be frozen instantaneously, stopping all biological processes within seconds and reducing the risks of contamination via handling. The method is widely used for extracting different chemical compounds, proteins, DNA, and RNA. Cellular liquids are frozen or crystallized during the freezing process, which causes intracellular structures to rupture, making this method not suited for morphological or ultrastructural studies.

4) **Dried biological material/herbaria**. Drying can be used as a long-term preservation method, and is the most commonly used long-term preservation method for seaweeds. After identification macroalgae should carefully dried using special acid-free cartons. Information about the individual, collection place, date, and name of identifier, should follow the preserved individual. Many characteristics can normally be examined from dried specimens, but for better observation parts of a dried seaweed can be rehydrated. In many cases it is also possible to obtain good DNA from herbarium specimens of seaweeds for later genetic identification of dried seaweeds, depending on the condition of the seaweeds before drying and the storage conditions in museums.

5) **RNAlater°** (Life Technologies, 2011): an aqueous, non-toxic tissue storage reagent that efficiently penetrates biological tissue and stabilizes cellular RNA. RNA is volatile and unstable compared to DNA, but RNA can be conserved for years if properly fixed with RNAlater. Typically, the procedure is to fix the sample for approximately one day at room temperature and then store at −20 °C.

6) **Oxidizing agent – osmium tetroxide (OsO_4):** a staining (fixative) agent used in transmission electron microscopy, TEM, (Kiernan, 2000) and scanning electron microscopy. The agent cross-links proteins and oxidizes unsaturated double bonds of fatty acids, stabilizing the doubled lipid membranes of the cells. In addition, osmium tetroxide provides excellent contrast to TEM-based ultrastructural analyses (Spector and Goldman, 2006). Osmium tetroxide is a highly toxic and volatile oxide of osmium and should always be handled with caution in a dedicated lab.

7) **Picrates:** includes fixatives with picric acid, commonly used in histology. It works well for soft tissue and provides the tissue with excellent contrast if stained with Hematoxylin and eosin. Note: the fixative does not work well for ultrastructural electron microscopy-based work.

Most commonly used is **Bouin solution** or "Bouin's fluid" (Bouin, 1897). Bouin's fluid is composed of picric acid, acetic acid, and formaldehyde in an aqueous solution. Each component has a specific function that complements the others. Acetic acid rapidly penetrates the tissue and causes protein denaturation and coagulation of nucleic acids. At a slower rate, formaldehyde molecules penetrate the tissue and bind proteins. Acetic acid causes the tissue to swell, but formaldehyde and picric acid counteract by causing moderate shrinkage. The slowly penetrating picric acid acts to coagulate and precipitate the proteins by forming salts with basic amino acids. The basic proteins are well preserved, while acidic proteins are not. Nuclear proteins also coagulate, but the DNA remains water-soluble and will, together with the acidic proteins, be removed during the tissue cleansing procedure following the fixation.

Picric acid is bright yellow and provides the characteristic color of Bouin's fluid and tissue exposed to it. Dry picric acid is explosive and sensitive to vibrations, so safety recommendations include storing picric acid wet and keeping it in plastic, not glass, bottles (Baker, 1958). A typical recipe for Bouin's fluid is: 75 ml picric acid (saturated aqueous solution), 25 ml 40% formalin (aqueous solution), and 5 ml glacial acetic acid.

8) **Freezing:** often done when detailed analyses (e.g., stomach analysis, parasite inspection) are needed and time, equipment, or personnel do not allow it to be completed in the field. Freezing samples may entail placing them in a freezer (−20 °C) or flash freezing (−196 °C) with liquid nitrogen. Flash freezing is used to minimize the size of ice crystals that form in tissues.

4.7.2   Health and Security When Using Fixatives

In general, all compounds used for tissue fixation and storage should be handled with caution. Fixatives are designed to immediately arrest autodegradation and enzymatic cell activities in the tissue, so that a snapshot of the tissue can be preserved for later studies. Consequently, fixatives are poisonous to living organisms. Fixation should always take place in a fume hood with proper ventilation and while wearing proper safety equipment (e.g., lab coat, eye protection (or keep the fume hood lowered to avoid splashes), and plastic gloves. These safety precautions should be strictly followed even when using a relatively mild fixative. Mild fixative, such as ethanol, must be applied in high concentrations to function as an efficient protein denaturing agent; these concentrations are hazardous to breathe and will, if coming into contact with the skin, cause damage. When dealing with chemical and buffers, *always* carefully read the manufacturer's instructions regarding storage and handling. Safety data sheets can be found at: http://ecoonline.com/search-sds/. Most institutes and laboratories will have safety protocols in place for each chemical – follow them! Safety sheets for fixatives are required to be brought onboard by the users.

Most of the primary and secondary fixatives and buffers listed in Section 4.7.1 require special attention and safety precautions. These include:

1) **Sodium cacodylate** (C2H6AsNaO2) (Kenyon and Hughes, 2001) is extremely toxic through ingestion, inhalation, or skin contact. Sodium cacodylate is carcinogenic and a tumor-promoting compound and should be treated with caution. Sodium cacodylic acid should always be handled under a fume hood. Wear disposable nitrile gloves (not latex) and a lab coat, when handling.

   Storage: mixtures of dissolved sodium cacodylate and water should be kept in a glass-stoppered reagent bottle, with the top wrapped with parafilm and stored in a refrigerator dedicated to chemical compounds.

   Disposal: Consult the local laboratory responsible.

2) **RNAlater°** is an aqueous, salt-based, non-toxic tissue storage reagent.

   Storage: RNAlater° can be kept in a sealed container with a screw cap at room temperature. Keep the container tightly closed when not in use because evaporation may occur. When fixing tissue in RNAlater°, keep the sample in the fridge overnight and then transfer to a freezer for long-term storage.

   Disposal: Consult the local laboratory responsible.

3) **Picric acid and picrate** (Mallinckrodt Baker Inc., 1998) is explosive if dry; store wet. Picrate can cause severe eye irritation and is harmful if swallowed, inhaled, or absorbed through the skin. Compounds containing picric acids should always be handled under a fume hood, using standard plastic gloves and eye protection (safety glasses).

   Storage: store in a bottle under water. Spontaneous explosion can occur if the picrate is allowed to dry in, for example, a pipette that has not been flushed with water after use.

   Disposal: Consult the local laboratory responsible.

4) **Aldehydes (formaldehyde and glutaraldehyde)** (Rasmussen, 1974) are suspected carcinogens and reproductive hazards. Work in a fume hood when using. Once aldehyde-fixed specimens have been transferred to ethanol or isopropanol, a fume hood is not strictly necessary.

   Storage: All aldehyde solutions must be stored in sealed, correctly labeled containers to prevent leaks and evaporation.

Disposal: As for the working solutions must be placed in sealed, correctly labeled containers. Spilled waste can be cleaned up using an absorbent material, which is then placed in sealed plastic bags. All waste must be stored for disposal by a certified disposal company. Do not dispose of waste in the trash or sink.

5) **Osmium tetroxide or OsO4** (Acros Organics N.V., 2008) is an extremely potent oxidizing compound and is highly toxic if inhaled or comes into contact with skin. Osmium tetroxide solutions must be prepared and handled in a certified chemical hood. Safety equipment includes a lab coat with sleeves rolled down, safety glasses, and double Nitrile gloves. Freshly prepared osmium tetroxide in a water-based solution is colorless to pale green in color, but turns black when oxidized; good to know in case of accidental skin exposure (black dots on skin).

Never work with Osmium tetroxide without permission and having consulted the local laboratory responsible.

Storage: Store in an area where only authorized personnel have access. A refrigerator containing osmium tetroxide must be isolated from the working area and carry a warning about the presence and risk of osmium tetroxide. Store solutions in a glass-stoppered, glass container inside a solid (plastic or glass) secondary container. Both containers must be labeled with the chemical name and a hazard warning.

Disposal: Solutions and solid waste material should be collected in a labeled, leak-proof container. All empty containers that were used to handle this chemical, for example, pipette tips, gloves, ampoules, must be placed in a rigid container. Disposals are handed to a hazardous waste pickup service. A completed waste label must be attached to the container.

6) **Liquid nitrogen** will cause cold burns if spilled on skin. Special gloves (not standard plastic laboratory gloves) should be used while handling.

Storage: Store liquid nitrogen in approved thermo containers. Eventually, it will evaporate, even in closed containers. Flash-frozen tissue should be stored below −20 °C.

7) **Alcohol (ethanol)**, in high concentrations, is hazardous to ingest and inhale; direct skin contact should be avoided. Handling must take place under a fume hood or in a well-ventilated area. The use of plastic gloves is advised.

Storage: Stored at room temperature in a tightly sealed container (plastic or glass). Ethanol-fixed tissue should be kept cold in a fridge or freezer.

References

Acros Organics N.V. (2008) Osmium(VIII)-Tetroxide, 99.9 + % (data sheet).
Baker, J.R. (1958) *Principles of Biological Microtechnique: A Study of Fixation and Dyeing*, Methuen & Co Ltd., London.
Bouin, P. (1897) Etudes sur l'évolution normale et l'involution du tube séminifère. *Archives d'anatomie microscopique*, 1, 225–339.

Kenyon, E.M. and Hughes, M.F. (2001) A concise review of the toxicity and carcinogenicity of dimethylarsinic acid. *Toxicology*, 160(1–3), 227–236. DOI:http://dx.doi.org/10.1016/S0300-483X(00)00458-3

Kiernan, J.A. (2000) Formaldehyde, formalin, paraformaldehyde and glutaraldehyde: what they are and what they do. *Microscopy Today*, 1(5), 8–12.

Kristoffersen, J.B. and Salvanes, A.G.V. (1998) Effects of formaldehyde and ethanol preservation on body and otoliths of *Maurolicus muelleri* and *Benthosema glaciale*. *Sarsia*, 83(2), 95–102. DOI:10.1080/00364827.1998.10413675

Life Technologies (2011) *RNAlater° Tissue Collection: RNA Stabilization Solution.* pp. 1–12.

Mallinckrodt Baker Inc. (1998) *Picric Acid, Wet.* (data sheet).

Melle, W., Ellertsen, B. and Skoldal H. (2004) Zooplankton: The link to higher trophic levels, in *The Norwegian Sea Ecosystem* (ed. H.R. Skjoldal). Tapir Academic Press, Trondheim, pp. 137–202.

Mjanger, H., Hestenes, K., Svendsen, B., Senneset, H. and Fotland, Å. (2016) *IMR Manual for sampling of fish and crustaceans.* Version 4.0 (SPD), August.

Rasmussen, K.E. (1974) Fixation in aldehydes a study on the influence of the fixative, buffer, and osmolarity upon the fixation of the rat retina. *Journal of Ultrastructure Research*, 46(1), 87–102. DOI:https://doi.org/10.1016/S0022-5320(74)80024-9

Spector, D.L. and Goldman, R.D. (2006) Constructing and expressing GFP fusion proteins. *Cold Spring Harbor Protocols*, 2006(7). DOI:10.1101/pdb.prot4649

Thavarajah, R., Mudimbaimannar, V., Elizabeth, J., Rao, U. and Ranganathan, K. (2012) Chemical and physical basics of routine formaldehyde fixation. *Journal of Oral and Maxillofacial Pathology*, 16(3), 400–405. DOI:10.4103/0973-029x.102496

5

Data Analysis

Knut Helge Jensen*, Jennifer Devine, Henrik Glenner,
Jon Thomassen Hestetun, Anne Gro Vea Salvanes and Kjersti Sjøtun

This book mainly focuses on using R for data analyses. R is a programming language for statistical computing and graphics (R project core team, 2016). There are several reasons why we have chosen the R environment for analyses and plots, the most important are listed below:

- It is a modern, object-oriented programming language that allows one to work efficiently and with high flexibility.
- It is a very popular statistical package among biologists and professional statisticians, making it easy to find help documentation using internet search engines, books, or by asking other users.
- Binary install files are available for all frequently used desktop platforms like Windows, MacOS, and Linux distributions (e.g., Debian, Ubuntu, Mint).
- It is open source software and downloadable for free at http://www.r-project.org/.
- Almost every analytical tool needed is available.
- It provides an interactive experience – one can modify and build functions to suit personal needs.
- Many people write new functions, so the capabilities of R are constantly expanding.

5.1 Scripts

The syntax (or commands or code) to perform actions in R can be written within the control console, but that way of working does not allow for easy replication of work. Working with a script (typing commands into a separate file) will ensure the exact steps (from import to data cleaning to analysis and plotting) can be repeated again, either with the same data set or, with slight modification, for other data. For this a text editor is needed; ideally, one that includes the R language so that syntax is highlighted (functions, arguments, parameters, and com-

---

* Lead authors; co-authors in alphabetical order.

Marine Ecological Field Methods: A Guide for Marine Biologists and Fisheries Scientists,
First Edition. Edited by Anne Gro Vea Salvanes, Jennifer Devine, Knut Helge Jensen,
Jon Thomassen Hestetun, Kjersti Sjøtun and Henrik Glenner.
© 2018 John Wiley & Sons Ltd. Published 2018 by John Wiley & Sons Ltd.

ments are uniquely color-coded so that the script is very easy to read). If working in RStudio, a text editor is included with the program. Otherwise, text editors tend to be platform specific and all have their advantages and disadvantages. Some common text editors for Windows include Notepad++, Tinn-R, Emacs; for Linux the choices are numerous and some popular ones include Gedit (with plugin RGedit), Geany, Notepadqq, and Kate, Mac users might explore TextEdit or Smultron. These editor suggestions are just a starting point.

Many of the examples in Chapter 5 have scripts included to read in data, create plots, and/or complete statistical analyses. Within the scripts, many notes or comments are included. One should always annotate one's scripts with comments as this will assist in remembering why a certain block of code was used (instead of using, e.g., a different function or method). Remember, while it may be easy to remember why something was done 1–2 weeks after completion, it is not so easy 1–2 months (or years) later.

R syntax within this chapter will be shown in **Courier New** font and bold typeface. Comments within the R syntax are marked as blue text and start with a hash symbol (#). Scripts are included with this chapter for the examples and give an idea of how one could script while working. Many excellent sources on programming (for efficiency and speed) exist, but this is not the purpose of this chapter. Here, we aim to lead a beginner R user through some common ways of handling data; it is not meant to provide an exhaustive list of R capabilities.

5.2 Setting the Working Directory

After opening and before beginning any work in R, one should set a working directory. This points R toward the directory where the scripts or data that will be imported into R are saved. We strongly advise creating a folder called rwork and within it, creating three additional folders: data, figures, scripts. Note: they must be spelled **exactly** as written here and the working directory must point to rwork for the included syntax and scripts to work properly.

The *rwork* directory containing the three subdirectories, *data, scripts*, and *figures*, are available at http://filer.uib.no/mnfa/mefm/.

To see where the default working directory is currently set in R, type:

```
getwd()
```

For most, the default working directory is not where the data for the analysis lie nor is it where we want to save the figures made in R. We reset this directory using the setwd command and typing the *path name* as follows:

```
setwd("path name")
```

A shortcut exists for obtaining the path to the new working directory; this shortcut does not work with RStudio. Open R, drag a text file (not a folder) from the working directory into R. Ignore the error message that results from this action. In the example below, a text file already in the rwork directory called *readme.txt* was dragged into the R workspace (Figure 5.1, first command line in red). The path name, excluding the file name, is then copied into the command as shown in the second red line.

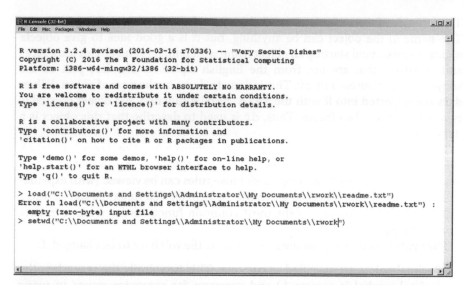

Figure 5.1 Example of how to obtain the path to a directory (first red line) and set a new working directory (second red line). Note that on MacOS or Linux, a forward slash instead of double or single backslash must be used. A forward slash will also work with Windows versions of R. A forward slash is recommended for producing platform independent syntax. Thus, a more general syntax for the second red line is: `setwd("C:/Documents and Settings/Administrator/My Documents/rwork")`.

Note! Should the working directory not be set to the `rwork` folder, the provided syntax and scripts will not work without modifying the path. R will remember the new working directory until R is closed. Every time R is reopened, the working directory must be reset.

5.3 Importing Data

Because R is an object-oriented programming language, objects must be created. The easiest way to create an object is to *read in* a data object. A data object can be in any form, but the most common are text files, where the file name ends in .txt or .csv (comma separated values).

In the example from the field work in Masfjord, western Norway as part of a course run by the Department of Biology, University of Bergen, data files have been saved in spreadsheets (see Chapter 4). R is capable of reading in files directly from some spreadsheet programs. However, we will keep things simple by using a single format, the comma separated values (.csv) file format. To convert a spreadsheet workbook into this format, first save each data sheet as a separate csv file. Once the data are saved as .csv files, the data files are imported into R using the `read.table` function:

```
zooplankton.df <- read.table('data/zoop.data.csv', sep=',')
```

With one line of syntax, data (*zoop.data.csv*) are imported and an object created, called `zooplankton.df`, which contains all the data that is in *zoop.data.csv*.

The left side of the *assignment symbol* (<-) is where the object is given a name. The name of the object can be anything, but it is a good idea to give objects logical names; avoid starting an object name with numbers and do not use spaces and/or letters that are not from the English alphabet. The last part of the zooplankton name ends in df. This is done to describe the type of object. When data are imported into R with the read.table function, the object created is typically called a data frame. Thus, df is used to describe that this object is a data frame.

The right side of the assignment symbol contains a function (read.table), the path, and the name of the csv file. Here, certain rules must be followed, which are defined by the function being used; those rules can be viewed (with explanation and examples on usage) by accessing the help for that function, e.g. ?read. table. We are using one of the most common functions for reading in data (read.table).

The syntax used to import the *zoop.data.csv* file will have to be changed if:

1) Decimal points are specified as commas. This is typically the case when the decimal symbol is comma (,) and common for computer setups in many European countries. Use the syntax:

```
zooplankton.df <- read.table('data/zoop.data.csv', dec=',', sep=';')
```

As we see, the sep statement changed because a format can't have the same symbol separating values as for decimal symbols. The best way to check which file format is used is to open the file in a simple text editor (not a word processor!) and take view at the data set.

2) Missing values are not coded 'NA', but as 'NULL' or left blank, use the following syntax, respectively:

```
zooplankton.df <- read.table('data/zoop.data.csv', na.strings='NULL')
zooplankton.df <- read.table('data/zoop.data.csv', na.strings='')
```

Thus, if missing values are left blank you specify this as a text string without any text within it.

3) Columns have names (or headers). Use the syntax:

```
zooplankton.df <- read.table('data/zoop.data.csv', sep=',', header=T)
```

4) If all of the above are true, use the syntax:

```
zooplankton.df <- read.table('data/zoop.data.csv',header=T, dec=',',
   sep=';',na.strings='NULL')
```

Remember that csv is a relatively standardized format. If data come in a different text file format, the solution is normally to open the file in a text editor and check the format. In this context it is worth mentioning that a tabulator (tab) is often used as separator between variables. The setting you use within the read. table function is then: sep='\t'.

If there were no problems with the data import, R will not give any message; the control prompt (>) will be waiting for the next command. If there was a problem, R *may not* give an error message. It is vital to check the imported data to verify the data have loaded properly. To do so, one or more of the following commands are useful: head(object   name), str(object name), names(object   name), which show the first six rows of data with variable names, the structure of variables in the data set, and the names of each variable in the data set, respectively:

```
head(zooplankton.df)
str(zooplankton.df)
names(zooplankton.df)
```

5.4   Working with Data

Once data has been read into R (Section 5.3), we are ready to begin to work with the data. All example data provided with this chapter are imported as data frames; data frames have rows and columns and all columns are of the same length. Each column is a *variable* in the data, for example, time of day, water depth, fish length, and actions (plotting, data analysis) are performed on these variables. There are two ways we can work with our variables.

1) Call each variable by first specifying the object name, $, then variable name:

```
zooplankton.df$dry.weight
```

This type of syntax is needed each time a variable is referenced. If only the variable name is typed into R (and data are not attached), an error message occurs, for example, Error: object 'dry.weight' not found.
2) Use the attach command to make variables available without having to specify the object name:

```
attach(zooplankton.df)
```

Actions can now be performed on variables within the object zooplankton. df by calling the variable name directly, for example, plot(dry.weight).

We do not recommend using the attach command for a data set that will be changed by creating new variables or correcting errors (Section 5.4.1). A data set that has been attached will not be updated with any changes unless it is reattached after each change. The syntax must then be full of attach commands and errors are easy to make if an attach command is forgotten.

5.4.1   Error Checking

Data sets can be large and data may be manually entered into the database from, for example, paper forms. Regardless of how carefully one works, there are always errors in data sets that need to be corrected before the analysis can begin. Preference may be to correct data sets by using "search and replace" in a spreadsheet. Before correcting the spreadsheet, we recommend saving a copy of the original, untouched file by putting, for example, "uncorrected" in front of the file name. Use the file without "uncorrected" in the name to correct errors and for subsequent analyses. Working in a spreadsheet is more self-explanatory than using a programming language like R and when first learning R, it may be faster (and easier) to work with the spreadsheet. However, R is much more efficient when working with larger data sets or for replicating a type of action.

We have created a small data set with some typical errors to show how to fix errors and save to a new, corrected csv file. The file with errors is called *faulty. data set.csv* and is found in rwork/data. The file contains the following errors:

1) Empty cells that are not coded as NA.
2) The decimal symbol is a period (.), but a comma (,) was used in one numerical observation. This is a serious mistake because the separator between variables in a csv file is also a comma. However, in spreadsheet programs like Excel and LibreOffice Calc, saving in standard csv file format will put quotation marks around such observations to mark them as categorical. This makes it easier to import the data set into R and correct the observation after import.
3) A letter was coded into the number in a numerical observation.
4) Names of some of the levels of categorical variables were misspelled.
5) One value of a variable is much higher than the others because a decimal point was omitted.

Normally, the errors in a data set are unknown. The syntax shown below is a general approach to find and correct errors like the ones listed above:

```
#Import the data:
faulty.df <- read.table('data/faulty.dataset.csv', header=T, sep=',')
```

When missing values are coded as blank cells in numerical variables, R will normally automatically correct these. For categorical variables and factors, missing values will remain blank. Correcting missing values can wait until other errors have been corrected.

```
#Check the structure of the data set:
str(faulty.df)
```

The output looks like this:

```
'data.frame':   6 obs. of  4 variables:
 $ Species: Factor w/ 3 levels "g.morhua","G.morhua",..: 2 2 1 3 3 3
 $ Sex    : Factor w/ 3 levels "female","male",..: 2 3 1 1 1 2
 $ Weight : Factor w/ 6 levels ""," 4","1","2.6r",..: 3 1 4 2 6 5
 $ Length : num   19.8 NA 497 60.3 NA 56.2
```

The output from the str function states that *Species* and *Sex* are factors. That is correct. However, *Weight* was imported as a factor, but it should be numerical. Typos in numbers are mistakes that commonly result in a numerical variable being coded as categorical; an exception is NA, which is the default code for missing values. These mistakes happen when any key other than a number or the decimal symbol was typed into a numerical variable, even if it is just in one observation. To search for this type of error in a variable, use the grep function:

```
grep("[a-z,*]", faulty.df$Weight)
```

The square brackets contain the symbols to search for, where the search specifies to look for any letter from the English alphabet a-z , then a comma, and finally a star. More characters can be added to the search string inside the square brackets.

The result from the above syntax

```
[1] 3 5
```

states that observations in row 3 and 5 match the search symbols. Look at rows three and five in the data set. Do this by extracting a subset of data from the data frame. Subsetting is performed using square brackets:

```
data.frame[i, j]
```

where i and j represent rows and columns, respectively. Here, we want to extract 2 rows (rows 3 and 5) and want all columns. Several rows can be selected using the c() function, which combines values. Thus, by writing c(3,5) in place of i in the syntax above and nothing after the comma, R will return 2 rows (3 and 5) and all columns in those rows. The syntax in our example therefore becomes:

```
faulty.df[c(3,5),]
```

The output looks like this:

```
    Species     Sex   Weight  Length
3 g.morhua   female    2.6r     497
5 P.virens   female     5,2      NA
```

The output shows two errors in *Weight*; a lowercase r has been typed at the end of the number in row 3 and a comma (,) was used instead of a period (.) for the decimal symbol in row 5.

We correct these two errors in two operations. However, the code can be written so that it is generic; that is, if an error was repeated several times, the syntax will correct it in one operation. Use the gsub function to correct the data, applying the following principle:

```
gsub('pattern', 'replacement', variable)
```

We use the gsub function to search for a pattern and when the pattern is found, R replaces that pattern with what is written in replacement. The variable tells R what variable to search for in the data frame for the pattern. Thus, for our data set, the syntax is:

```
faulty.df$Weight <- gsub('r', '', faulty.df$Weight)
faulty.df$Weight <- gsub(',', '.', faulty.df$Weight)
```

To the right of the assignment symbol, faulty.df$Weight tells R to put the corrections into the original data set, into the *Weight* variable.

Remember that *Weight* was imported as a factor, but it should be numerical. *Weight* must now be converted to a numerical variable. If the variable still contains non-numerical symbols in some observations, further error correction is needed before continuing, otherwise the syntax below will convert any numerical variable with characters to NA. Always check that the variable contains only numbers by looking at all the values by printing it to screen

```
faulty.df$Weight
```

Alternatively, if the data are many, use the syntax:

```
table(faulty.df$Weight)
```

If no characters remain, convert *Weight* to numerical by:

```
faulty.df$Weight <- as.numeric(faulty.df$Weight)
```

Check the structure of the data frame again to get an overview and verify that numeric variables are listed as numeric and categorical are characters or factors:

```
str(faulty.df)
```

The command should give the following output:

```
'data.frame':      6 obs. of  4 variables:
$ Species: Factor w/ 3 levels "g.morhua","G.morhua",..: 2 2 1 3 3 3
$ Sex    : Factor w/ 3 levels "female","male",..: 2 3 1 1 1 2
$ Weight : num  1 NA 2.6 4 5.2 3.7
$ Length : num  19.8 NA 497 60.3 NA 56.2
```

Weight and *Length* are now numerical (as seen from "num" in the output). The output also shows NA in *Weight*. By converting the variable to numeric, the original missing value in *Weight* is now coded as NA automatically.

Use another quick check of the data to verify the range (minimum and maximum values), or compare these values to the mean values. Use the summary function on the data object:

```
summary(faulty.df)
  Species        Sex         Weight          Length
g.morhua:1   female:3   Min.   :1.0    Min.   : 19.80
G.morhua:2   male  :2   1st Qu.:2.6    1st Qu.: 47.10
P.virens:3   Male  :1   Median :3.7    Median : 58.25
                        Mean   :3.3    Mean   :158.32
                        3rd Qu.:4.0    3rd Qu.:169.47
                        Max.   :5.2    Max.   :497.00
                        NA's   :1      NA's   :2
```

The variable *Weight* has minimum, mean, and maximum weights of 1.0, 3.3, and 5.2 kg, respectively; these look reasonable for Atlantic cod. The corresponding values for *Length*, on the other hand, are 19.8, 158.3 and 497.0 cm, respectively. One of the fish is nearly 5 meters long! An assumption might be that 497 should be 49.7, assuming that a comma has been forgotten or that a 1 was mistyped as a 4 when entering the data. One way to check is to compare the body mass to length by plotting *Weight* against *Length* as follows:

```
plot(faulty.df$Weight, faulty.df$Length)
```

The plot shows that the 497 cm fish is about 2.5 kg, so it makes perfect sense to replace 497 cm with 49.7 cm; the fish should not be 497 cm long because the weight is too light for that length. Replace the value using the sub command:

```
faulty.df$Length <- as.numeric(sub('497', '49.7', faulty.df$Length))
```

Note the difference between the two replacement commands used; sub (performs a replacement of the first occurrence of the match) and gsub replaces all matches within the variable.

Use the `plot` and `summary` functions again to see if there are any more odd values that need to be inspected and/or corrected. Continue to do so until no more errors are found.

From the summary output (`summary(faulty.df)`), *Species* shows three different levels: `P.virens`, `g.morhua`, and `G.morhua`, but two levels are actually the same species; Atlantic cod is both `g.morhua` and `G.morhua` in the data. The correct text is `G.morhua`. *Sex* also has a typing error; male is written both as `male` and `Male`. A better method for checking for misspelled words in a categorical variable is to use the `levels` function:

```
levels(faulty.df$Species)
```

The output from this command is:

```
"g.morhua" "G.morhua" "P.virens"
```

The `gsub` command can be used to correct all misspellings within a level of a categorical variable. The `as.factor` function is also applied on the categorical variable in the same operation. This is done to ensure that the categorical variables are factors, which is needed for the subsequent analyses, and does not become a character variable. An Internet search on *"R pitfall" + factors* can give more information on this topic.

```
faulty.df$Species <- as.factor(gsub('g.morhua', 'G.morhua',
    faulty.df$Species))
faulty.df$Sex <- as.factor(gsub('Male', 'male', faulty.df$Sex))
```

Using the `summary` function on the data frame or the `levels` function on each of the two variables should show they are now correct.

5.4.2 Saving Data

Once finished with error correction, the corrected data frame is ready to be saved as a csv file; this means the correction process is not needed each time the file is imported. Save the file using the `write.csv` function, where `file =` points at the subdirectory of `rwork` (e.g., `rwork/data/`, where the original data set called *faulty.dataset.csv* was saved):

```
write.csv(faulty.df, file='data/corrected.dataset.csv', row.names=F,
    quote=F)
```

The argument `row.names=F` is written to avoid R creating a name for each row of the data frame, while `quote=F` means quotation marks will not be around each field. If a csv file is wanted that has a decimal symbol as a comma (,) and separator as a semi-colon (;), use the `write.csv2` function instead of `write.csv`.

To summarize this section, error correction within R consists of two parts: searching for errors and fixing the errors found in each search. To search for errors, we have shown how to utilize the following functions: `str`, `grep`, `summary`, `levels`, and `plot`. In addition, how to subset rows and columns in a data frame was learned. To fix the errors, the functions; `gsub`, `sub`, `as.numeric`, and `as.factor` were shown. When fixing errors in a data set, there is no need to find all the errors before fixing them; instead, search and fix what was found, then repeat, until no more errors are found.

5.5   Data Exploration and Statistical Testing

The chapter begins with a short introduction to multivariate statistics, which are used for questions where there are many response variables, before moving to univariate statistics, which involve only one response variable at a time. All of the examples will be from the Masfjord example data and will explore specific questions or hypothesis testing related to those data. As the chapter progresses, less detail on the R syntax will be provided, especially if functions and arguments have been defined previously. Before any statistical analysis, we generally recommend to plot the data every way imaginable; this will ensure a full overview of the data and may lead to additional questions for analyses.

5.5.1   Analysis of Marine Communities

Multivariate statistics are often used in quantitative community ecology, when an overview of the species community is needed. These methods are particularly useful when the studied species are relatively stationary, for instance like benthic flora and fauna; however, these methods are not limited to these types of communities. Ordination methods are an often used type of multivariate statistic. All ordination methods aim to reduce dimensions so that the variability of the data can be explained through a few axes or gradients. Ordination can be divided into two categories called indirect and direct ordination, although this is a coarse generalization. Indirect ordination does not involve hypothesis testing, while direct does. Direct methods ordinate the data according to one or more explanatory variables (e.g. temperature, salinity, wave exposure) and then determine how the response variables correlate to the ordination scores. Indirect methods ordinate response variables (against themselves).

At the time of writing, one of the best R libraries for performing both types of ordination methods is called *vegan* (Oksanen *et al.*, 2016). This package is not a part of the base R installation and must therefore be installed as an extra package. To install extra packages in R is simple. From within R, write the following:

```
install.packages('name.of.library')
```

To install the *vegan* library, type:

```
install.packages('vegan')
```

A box may open that shows R repositories from around the world; choose one.

Data sets used for ordination do not have a header name for the data in the first column; this column is used for row names that represent the names of sites that are sampled. Thus, a typical structure of a data set prepared for ordination looks like the one shown in Table 5.1.

One of the most frequently used indirect ordination methods is called Principal Component Analysis (PCA). A simple PCA on the above data is done as follows:

```
#Import the data
pcadata.df <- read.table('data/pca.exampledata.csv',header=T,
  row.names=1, sep=',')
attach(pcadata.df)
```

Table 5.1 Typical structure of a multivariate data set prepared for ordination analysis. Note that the first column lacks a header name. This column represents row names, which are the different sites investigated. The numbers in each column represent percent cover of a given species within a given site.

	SpeciesA	SpeciesB	SpeciesC	SpeciesD	SpeciesE
Site1	46	80	66	24	94
Site2	38	86	44	51	91
Site3	16	89	65	78	92
Site4	81	15	31	97	13
Site5	37	15	96	94	17
Site6	89	5	97	25	31

Note! A new syntax is used in the import command compared to what was learned in Section 5.3. For PCA the import syntax must include row.names=1. This argument tells R that the first column in the data set is not a variable, but represents row names.

```
#Load the required library
library(vegan) #assumes that vegan is already installed. If not,
   run the install.packages command given above.

#Perform the analysis
pca1 <- rda(pcadata.df)
biplot(pca1)
```

The resulting plot is interpreted as explained in Figure 5.2.

Ordination is all about reducing dimensions; knowing how much of the variability of the data is explained by each axis is important. Eigenvalues contain this information. Eigenvalues measure the amount of variation explained by each principal component (PC) and will be largest for the first PC and smaller for subsequent PCs. Ideally, the variation in the data should be explained with few axes, that is, 2–3 axes represent 70–80% of the variation in the data. To get the eigenvalues, use the summary function of the PCA:

```
summary(pca1)
```

The main results are shown below. In this example, only two axes are needed to explain approximately 88% of the variation in the data, as seen from the cumulative proportion values.

```
Eigenvalues, and their contribution to the variance
Importance of components:
                         PC1       PC2      PC3      PC4      PC5
Eigenvalue            3772.5827 1327.5954 709.6669 10.11916 0.13587
Proportion Explained    0.6482    0.2281   0.1219  0.00174 0.00002
Cumulative Proportion   0.6482    0.8763   0.9982  0.99998 1.00000
```

An important assumption must be met to use PCA; the response curve must be linear. Sites are assumed to represent an environmental gradient and the response

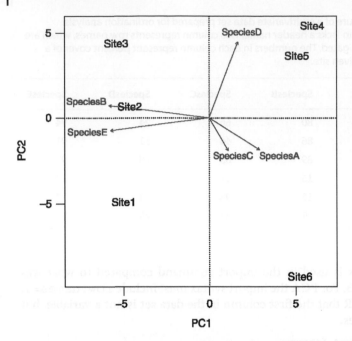

Figure 5.2 A PCA analysis plot from the data shown in Table 5.1. PCA plots are interpreted as follows: sites that are close together in the diagram have a similar species composition; sites 4 and 5 are quite similar. The origin (0,0) is species averages. Points near the origin are either average or poorly explained. Species increase in the direction of the arrow and decrease in the opposite direction. Distance from the origin reflects the magnitude of change. Variables near each other are similar. Angles between arrows approximate their correlations: 90° = 0 correlation, <90° = positive correlation, >90° = negative correlation, and 0° implies correlation = 1.

curve must increase (or decrease) linearly over the sites within the investigated gradient. To test this is true, use the decorana function on the same data and check axis lengths.

```
#Check axis lengths
decorana(pcadata.df)
```

The general rule for how to decide about axis lengths is where gradient length is:

- >4: unimodal response curve;
- 3–4: probably unimodal;
- 2–3: probably linear;
- <2: linear response curve.

From the data shown in Table 5.1, maximum axis length was 1.011 and the response curve can be assumed to be linear.

In general, use the two categories of ordination (direct or indirect) and information about whether the response curve is linear or unimodal to determine the kind of ordination to use in the analysis (Table 5.2). With the above example, if the result had indicated a unimodal response curve, a detrended correspondence analysis (DCA) instead of a PCA should be used.

Table 5.2 A key for how to choose which method of ordination to use.

Ordination	Response Model	
	Linear response	Unimodal response
Indirect	Principal Component Analysis (PCA)	Correspondence Analysis (CA) Detrended CA (DCA)
Direct	Redundancy Analysis (RDA)	Canonical Correspondence Analysis (CCA)

The functions to use from the *vegan* library of R to perform the different analyses listed in Table 5.2 are: rda for PCA and RDA, cca for CA and CCA and decorana for DCA. Be aware that the interpretation of plots depends on type of ordination, that is, the interpretation principles of the PCA plot in Figure 5.2 cannot be used to interpret a DCA.

When performing ordination, avoid the arch effect and the horseshoe effect. Both are explained at http://ordination.okstate.edu/glossary.htm. Recommended further reading about ordination methods are http://ordination.okstate.edu/ and references therein. The two documents written made for the *vegan* library also provide background information and include the *vegan* tutorial: http://cc.oulu.fi/~jarioksa/opetus/metodi/vegantutor.pdf and the introduction to the library: https://cran.r-project.org/web/packages/vegan/vignettes/intro-vegan.pdf.

5.5.1.1 The Bray-Curtis Dissimilarity Index

The Bray-Curtis dissimilarity index (Bray and Curtis, 1957) is a popular tool to use in Multidimensional Scaling (MDS) and cluster analysis dendrograms for answering questions like: do invertebrate communities (or assemblages) differ between sites? First, we illustrate how the Bray-Curtis dissimilarity index is calculated manually. We use R to perform the calculations and MDS and then make a cluster analysis dendrogram.

The Bray-Curtis dissimilarity index is a comparative measure between two variables, each with a column of associated data. In the example data below (Table 5.3), the Bray-Curtis formula (Equation 5.1) is used to estimate an index value representing the degree of dissimilarity between two data sets, which ranges between 0 (completely similar) and 1 (completely dissimilar). The data are the number of individuals or % cover of each alga species at two stations, where the data are from one square quadrat plot at each station. This is done as follows:

For each row for both stations, take the sum of the *absolute difference* between the row entries and divide it by the *total sum* of the row entries (Table 5.3; Equation 5.1).

$$D_{jk} = \frac{\sum_{i=1}^{a} |Y_{ij} - Y_{ik}|}{\sum_{i=1}^{a} (Y_{ij} + Y_{ik})}$$

(5.1)

Equation 5.1   The Bray-Curtis dissimilarity (D_{jk}) comparing stations j and k for all species (a). Y_{ij} is the amount (number or % cover) of species i at station j, Y_{ik} is the amount (number or % cover) of species i at station k.

Two different types of data exist in this data set: number of individuals and % cover. The values of these two data types are different, one is an absolute number and the other is a percentage. Large values (high % cover or high numbers of animals) will be disproportionately influential if the Bray-Curtis formula is applied directly to the data, as was done above. A data transformation is needed to make the values more similar to each other; that is, apply a formula to each cell in the table to homogenize the values (or to place values on a similar scale). Use a fourth root transformation ($x^{1/4}$), which means that the fourth root of each cell value is used to create a table of transformed values that are then used to calculate the Bray-Curtis dissimilarity value (Table 5.4).

Table 5.3 An example of applying the Bray-Curtis dissimilarity index to a hypothetical case with samples from one square quadrat plot at each station. According to Equation 5.1, $D_{jk} = 140/(165+145) = 0.45$.

Species	Alga a (%)	Alga b (%)	Alga c (%)	Animal 1 (no.)	Animal 2 (no.)	Sum
Station 1	20	50	90	0	5	165
Station 2	10	85	20	10	20	145
Abs. diff.	10	35	70	10	15	140

Table 5.4 The values of Table 5.3 after a fourth root transformation. This homogenizes values to give a more accurate picture of differences between the stations. Recalculation of D_{jk} changes it to 0.20.

Species	Alga a (%)	Alga b (%)	Alga c (%)	Animal 1 (no.)	Animal 2 (no.)	Sum
Station 1	2.114743	2.659148	3.08007	0	1.495349	9.34931
Station 2	1.778279	3.03637	2.114743	1.778279	2.114743	10.82241
Abs. diff.	0.336464	0.377222	0.965327	1.778279	0.619394	4.076686

In the next example, we investigate whether habitats that differ in wave exposure also tend to differ in species composition using the data set, *littoraldata.csv*. The data represent a design with four littoral zone sites: two with direct wave exposure and two that are located in the same area but are more sheltered from waves. For each site, there are four subsamples. A Bray-Curtis dissimilarity index is applied to make pairwise comparisons of all the subsamples. These values are used to generate a Bray-Curtis dissimilarity matrix, shown in Table 5.5.

This matrix by itself is difficult to interpret, but the information can be visualized for easier interpretation. We will use two such methods: a cluster dendrogram and a MDS plot. A dendrogram (Greek: dendron, tree and -gram, a written record) is a graphical means to visualize a hierarchical cluster analysis. Different methods exist to calculate clustering; in this example, the method used is "group average". Similarity values are presented in a tree structure, the branches group successive clusters of Bray-Curtis dissimilarity values. The degree of dissimilarity is plotted

Table 5.5 A Bray-Curtis dissimilarity matrix showing the calculated pairwise dissimilarities between samples taken at four sites (four subsamples at each site), where two of the sites are sheltered from waves (s) and the other two are exposed (e).

	11e	12e	13e	14e	21e	22e	23e	24e	31s	32s	33s	34s	41s	42s	43s	44s
11e																
12e	0.39															
13e	0.34	0.23														
14e	0.51	0.35	0.21													
21e	0.27	0.20	0.15	0.31												
22e	0.36	0.17	0.16	0.20	0.16											
23e	0.35	0.27	0.21	0.29	0.11	0.18										
24e	0.32	0.21	0.13	0.22	0.10	0.12	0.08									
31s	0.73	0.85	0.87	0.89	0.87	0.87	0.88	0.88								
32s	0.66	0.76	0.78	0.81	0.78	0.78	0.79	0.79	0.48							
33s	0.68	0.78	0.71	0.74	0.71	0.70	0.72	0.72	0.48	0.30						
34s	0.73	0.83	0.75	0.79	0.75	0.75	0.76	0.76	0.72	0.76	0.46					
41s	1.00	1.00	1.00	1.00	1.00	1.00	1.00	1.00	0.58	1.00	1.00	1.00				
42s	0.74	0.87	0.88	0.90	0.88	0.88	0.89	0.89	0.53	0.38	0.57	0.70	1.00			
43s	0.65	0.65	0.69	0.75	0.70	0.69	0.72	0.71	0.59	0.45	0.31	0.61	1.00	0.43		
44s	0.86	0.86	0.88	0.90	0.88	0.88	0.89	0.88	0.76	0.74	0.62	0.31	1.00	0.54	0.48	

on the Y-axis and the sites or subsamples are on the X-axis. In a MDS plot, the distance matrix is used to compute a set of vectors, then the Bray-Curtis values are displayed as points on a (usually) two-dimensional plane. Points closer to each other are more similar.

The following five-step guide outlines how to analyze a data set containing species data from quadrat plots into Bray-Curtis values, then visualizing the data using a cluster dendrogram and a MDS plot. Manual calculations are not needed because R will perform all the calculations. Remember to set the R working directory to the directory containing the data, scripts, and figures sub-folders, that is, set the working directory to rwork.

1) *Preparing the species data frame for import to R*

   Prepare a spreadsheet containing species abundance data from several sites and subsamples within each site, where site (and subsample) identifiers are in the first column and species abundance data for each site and subsample are in subsequent columns. The first column should not contain a header name. Missing observations should never be coded as empty cells, but as NA. Sites that did not record a particular species should be coded as zero (0). Logical names must be made for each site to understand the plots that will be created (Table 5.6). Here, the first number in the row names represents site ID, the second number represents the subsample number within the site (four sub-samples within each site), and the letter represents site type (e for exposed,

Table 5.6 Part of the example data set showing sites as row names (first column), where the first and second numbers represent site and subsample number, respectively. The letter at the end on each row name represents site type, where "e" means wave exposed and "s" is sheltered site. The rest of the data are the species abundance matrix.

	Ascophyllum nodosum	Fucus spiralis	Fucus vesiculosus
11e	0	0	0
12e	0	0	16
13e	0	0	4
14e	0	0	28
21e	0	0	8
22e	0	0	32
23e	0	0	64
24e	0	0	36
312	0	3	0
32 s	0	93	0
33 s	8	38	0

s for sheltered); write documentation to explain the row IDs. If the analysis is later expanded to one of the direct ordination methods mentioned in Table 5.2, linking to an environmental variable later will then be easy.

Once this spreadsheet is made, save it to import into R. In a spreadsheet program, choose *File*, *Save As*, then chose "CSV (Comma Separated Values, *.csv) as the file type. Call the file *littoraldata.csv* to ensure the syntax given in the following section works with the data set. Ignore any warnings regarding incompatible features. Be sure to save the file in the data subfolder of rwork.

2) *Importing the data into R*
Assuming that the data set was saved using the same name as the training data set, *littoraldata.csv*, import it as follows:

```
littoral.df <- read.table('data/littoraldata.csv', header=T, row.names=1,
   sep=',')
```

Any errors in importing the data may be due to the format. A non-English computer setup likely uses a semi-colon to separate variables in the csv file instead of a comma and the decimal symbol is a comma instead of a period. The easiest way to check this is to open the csv file in a text editor (not a word processor!) like Notepad in Windows and look at the format. How to import a csv data file saved from a non-English computer setup was explained in Section 5.3.

3) *Loading the vegan library for multivariate statistics*
The analysis will require a function from the *vegan* library, therefore, load the *vegan* library into the R workspace:

```
library(vegan)
```

4) *Producing a Bray-Curtis dissimilarity matrix, cluster dendogram and MDS plot*
Create the Bray-Curtis dissimilarity matrix, similar to that shown in Table 5.5, the cluster dendogram, and MDS plot, by using a function found in `rwork/scripts/functions`. If the working directory of R is set to the `rwork` directory, that is, by using `setwd(path name)`, load the function into R by:

```
source('scripts/functions/cluster.r')
```

Use the `cluster` function that was put into the R workspace with the source function:

```
clusterplots(dataset, group)
```

where `dataset` is the name of the data frame created in step 2 and `group` is a threshold level for the groups within the plots. This value should be set to 0.41 as a starting point.

The function can be used as follows:

```
clusterplots(littoral.df, 0.41)
```

If a warning message similar to "`the matrix is either rank-deficient or indefinite`" appears, the MDS plot most likely contains a "cluster" formed by a straight line. If so, ignore the warning.

The function may need slight modifications with each new data set. Modifications may include changing the colors of the points (line 41 in the function script called cluster.r), renaming the plot titles (line 44), or placement of the legend (line 47). These changes are done by opening the function script, making the modifications in the correct place, and saving the script; the function must be reloaded into R after each time it is modified and saved. A function may be loaded as many times as needed during an R session. The last load of the function is the one that R will use.

5) *Checking the results*
Check the figures to see if they make sense. Consider whether the number and threshold level of the groups within the analyses are appropriate for the data; if not, change the *group* value for the clusterplots function and run it again. In the training data set, the variability between subsamples within the same site is quite large, so the value of the *group* variable can be increased from 0.41 to 0.6 as follows:

```
clusterplots(littoral.df, 0.6)
```

The resulting plots are shown in Figure 5.3 and 5.4.

5.5.2 Physical Environment

Environmental data should always be included when evaluating distributions of different species. A CTD is commonly used to measure environmental data. Although CTD is an abbreviation for Conductivity, Temperature and Depth, modern CTD units often have possibilities to measure more than these three parameters. We have made a function that can be used to make temperature,

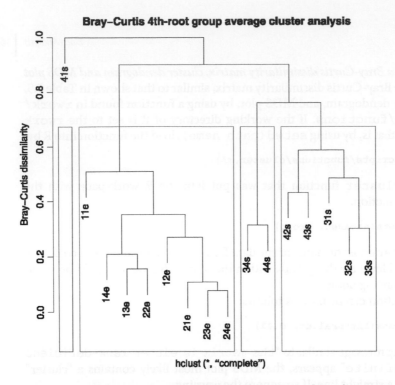

Figure 5.3 A dendrogram showing a hierarchical clustering of sampling sites 11e–24e (samples from wave exposed sites) and 31 s–44 s (samples from sheltered sites) based on a Bray-Curtis dissimilarity matrix. Connections made at values close to 0 indicate sample sites have a high degree of similarity ("low degree of dissimilarity").

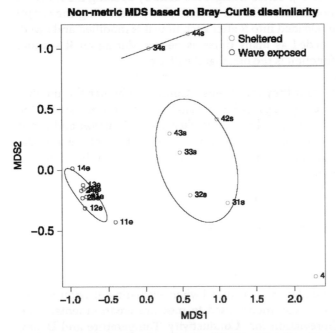

Figure 5.4 A MDS plot showing distances between sampling sites 11e–24e (samples from wave exposed sites) and 31 s–44 s (samples from sheltered sites) on a two-dimensional plane. The elliptical borders show cluster overlays from the separate cluster analysis with dissimilarity less than 0.6. For 34 s and 44 s the ellipse has collapsed since there are only two data points in this cluster.

oxygen, salinity, and light profile plots from standard ASCII file CTD output. Load the function into the R workspace by using the source command:

```
source('scripts/functions/CTD.plot.r')
```

and then use the function as in the following example:

```
CTDplot(Sta0683)
```

where Sta0683 represents the file name of the cnv file containing the data. These cnv files are created after converting the downloaded CTD data from the CTD unit using (in most cases) the standard conversion software provided by the company. Note that the file name in the function does not include *.cnv*. The plot from this function is shown in Figure 5.5.

Many ways exist to display CTD data. The oce package can be used (Kelley *et al.*, 2016). More info can be found here: https://cran.r-project.org/web/packages/oce/oce.pdf.

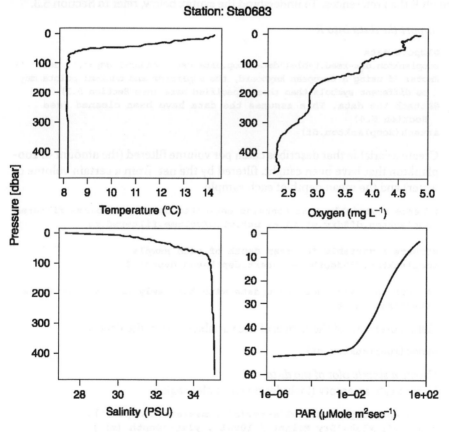

Figure 5.5 Environmental data from CTD station 0683. The panels show profiles of temperature, oxygen, salinity, and light (PAR) over depth (m). The CTD records pressure (dbar) which is then used as the depth. Note that the depth is not on the same scale for the PAR curve because the light measuring unit cannot measure PAR values $\leq 10^{-12}$.

5.5.3 Zooplankton Samples

This section will detail how to use R to test a specific hypothesis on zooplankton data that was collected during field work in Masfjord. Chapter 4 illustrated how zooplankton samples could be split, where one fraction was used for species identification and other was for total biomass. Total biomass data will be used for the analyses in this section.

When testing hypotheses, it is always H_0 that is tested; the null hypothesis states that there is no effect of the predictors being tested. A hypothesis for the zooplankton samples may therefore be formulated as:

> H_0: *The depth distribution of zooplankton is not dependent on time of day (day or night).*

To answer this, we first begin by importing the data into R. As stated previously, the location of the file containing the data depends on where it was saved; however, we have stressed the importance of placing all data into one working directory, which R then references. To understand the syntax below, refer to Section 5.3.

1) *Import the data into R*

```
#Import data
zooplankton.df<-read.table('data/zoop.data.csv',header=T,sep=',', dec='.')
#Note! If using a European keyboard, the separator and decimal points may
  be different symbols than those specified here (see Section 5.3)
#Attach the data. This assumes the data have been cleaned (see
  Section 5.4)
attach(zooplankton.df)
```

Create a variable that describes catch per volume filtered (the amount of zooplankton that have been caught, filtered by the net, from a certain volume of water) and the mean depth of each sample.

```
#Create a variable that contains estimated catch per volume filtered
zooplankton.df$dry.wt.m3 <- net.wt.g/volume.filtered.m3

#Create a variable for mean depth of each sample
zooplankton.df$depth <- (start.depth+end.depth)/2

#Attach the data again to update with the newly created variables
attach(zooplankton.df)
```

Get an overview of the names of the variables within the data set:

```
names(zooplankton.df)
```

2) *Create a simple plot of the data*
First, create an empty (includes no data) plot region:

```
plot(depth~dry.wt.m3, ylim=rev(c(0, max(start.depth))),
type='n', xlab='Dry Weight / 1000L', ylab='Depth (m)')
```

The syntax appears quite complicated, but is very straightforward. The plot function creates the plot and must have several arguments passed to it (access help or view the arguments this function uses by typing ?plot). The first of these are instruction on what variables to plot, given here in the form of y~x (depth ~ dry.wt.m3). Limits for the y-axis are defined using the argument ylim and contain two values, 0 and the maximum start depth (the maximum value is found using the max() function). The values are combined using the c() function (see Section 5.4.1) and their order is reversed using the rev() function. The next argument type='n' tells R 'do not plot this figure'; this is used to set up the plot region (i.e., create and format the region as wanted, but not plot the actual data). The labels for the x- and y-axes are then defined using the xlab and ylab arguments. The graph is then populated with data using the points function:

```
points(depth[day.or.night=='day']+5~dry.wt.m3[day.or.night=='day'])
points(depth[day.or.night=='night']-5~dry.wt.m3[day.or.night=='night'], pch=19)
```

The first line plots data where time of day = 'day', while the second line plots only the observations for 'night'; points is used to specify the data are plotted as symbols, and not, for example, as lines. As with the plot function, instruction is needed on what data to plot, given here in the form of y~x (depth ~ dry.wt.m3). The information in the square brackets are the instructions to subset (see Section 5.4.1) based on time of day. The depth for each time of day category has been offset slightly (+5 for day, -5 for night) so that data points do not lie on top of each other (may also use jitter). The argument pch in the second line specifies the type of symbol to use; pch=19 is a filled circle (default fill color is black). The default symbol used when plotting is the open circle (used for day). This plot is shown in Figure 5.6.

3) *Save the plot*

```
#Save the plot
savePlot('figures/zooplankton.simple.png', type='png')
#Alternative method if savePlot does not work under Mac OS X
quartz.save('figures/zooplankton.advanced.png', type='png')
```

4) *Data analysis*
To test the null hypothesis:

> H_0: *The depth distribution of zooplankton is not dependent on time of day (day or night)*,

use a linear mixed-effects model (LME). This is done in R with the lme function from the nlme library (Pinheiro and Bates, 2000). Mixed-effects modeling is necessary because of the clustering of data within sampling sites (see Zuur *et al.*, 2009 for details on statistical methods and explanation of random effects). See Box 5.1 for details on the syntax written below.

```
#Perform the statistical analysis:
library(nlme) #To load required functions for LME models
fit1.lme <- lme(dry.wt.m3~factor(depth)*day.or.night, random=~+1|
    Station, data=zooplankton.df)
```

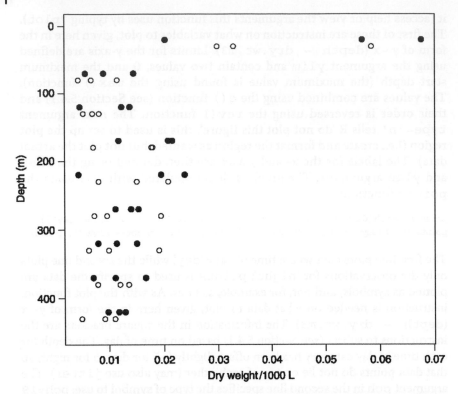

Figure 5.6 The density of zooplankton (measured in dry weight per cubic meter) at different depths and time of day, day and night. This figure was made using the code in *(2) Create a simple plot of the data*. Open and solid circles represent day and night samples respectively.

Once the model has run, view a condensed summary of the output by:

```
anova(fit1.lme)
                            numDF denDF  F-value  p-value
(Intercept)                    1    32 166.58351  <.0001
factor(depth)                  8    32  10.59350  <.0001
day.or.night                   1     4   1.16798  0.3406
factor(depth):day.or.night     8    32   0.12218  0.9979
```

The first p-value to view belongs to the interaction term of the model, that is, the last reported p-value, associated with factor(depth):day.or.night; this is 0.9979. The null hypothesis is rejected when p ≤0.05 (note: the significance level chosen for determining when to reject the null hypothesis is set by the researcher, but common values include 0.1, 0.05, and 0.01). From this p-value, we can conclude that the depth distribution of zooplankton is NOT significantly different between day and night. If this interaction was significant, the change in the amount of zooplankton with depth was different between day and night.

Because the interaction was not significant, next look at the line above it in the anova output (the effect of time of day on mean biomass of zooplankton). This is not significant, therefore the interpretation is that the overall biomass of zooplankton sampled during the day is not different from the amount sampled at night. Had it been significant, the interpretation would have been that the biomass of zooplankton sampled was significantly higher in one period than the other.

Box 5.1 The linear mixed-effects model syntax explained

fit1.lme is the name given to the model object. Any name can be used, but avoid starting an object name with numbers; avoid using spaces and/or letters that are not from the English alphabet.

<- is the assignment symbol in R. Because R is an object-oriented programming language, the arrow points to a named object (given on the left side) and puts everything on the other side of the arrow (the right side) into the object (in this example, the object is called **fit1.lme**). Thus, instead of writing all that is written on the right hand side of the arrow each time the results are wanted, for example, for plotting, only the object name needs to be called. Calling the last part of the object name something related to the type of object is a good practice. By doing so, it is easy to have an overview of the type of all objects created in the workspace (e.g., data frames, linear models, linear mixed-effects models). Using the suffix **.lme** informs us that the object is a fitted model object of a linear mixed-effects model.

lme(...) is a generic function that fits a linear mixed-effects model. The response variable (e.g., y, the variable of interest, dependent variable) is written to the left side of the tilde symbol (~), while the predictor(s) is(are) on the right side. The predictor variables (e.g., x, explanatory variables, independent variables) are the variables that are thought to predict the response variable. In the model, we test to see if the predictor variables explain a significant proportion of the variability observed in the response variable. The names of response and predictor variables must be as written in the data. In this example, the data are called *zooplankton.df*. To view variable names, write **names(name.of.data.frame)**, which in this case is **names(zooplankton.df)**.

factor is a function that has been called within the lme function. This specifies that the variable *start.depth* be treated as a factor in the model and not as a continuous numerical variable. Each variable is assigned a category when imported into the workspace. In this example, *start.depth* was imported as a numerical variable, but we want it to be a categorical variable (a factor). We do this because we expect that the catch of zooplankton does not follow a linear response with depth, but rather that it goes up and down over increasing depth. In our data, zooplankton have been sampled from discrete depth ranges, with multiple samples from a depth range, which is why we can specify depth as a factor.

* (multiplication sign) used between two predictors means that the interaction between the two variables will also be tested. In a model with two predictors A and B, writing A*B is the same as writing A+B+A:B. The multiplication sign is a shortcut. If testing the interaction is not wanted, write only A+B.

random is an argument specifying the random effect factor(s). In this example, the intercept of each station is a random effect, that is, the mean catch within each station is a random effect. For more details on random effects, see Zuur *et al.*, 2009.

data=specifies which data the model will use.

The results of the anova indicate that depth is significant; we interpret this as the amount of zooplankton depends on depth. This result does not give any indication of what depth had higher biomass than the other sampled depths.

In general, do not look at only the p-values; p-values are meaningless without information about the magnitude or direction of the effect. This information can be gathered from the summary output of the model (use `summary(model object name)`) and by plotting the data.

5) *Summary output: view and interpret*

We have made a script that can be used to make a plot of zooplankton biomass by depth and time of day (Figure 5.7). Again, the script can be opened in any editor program and run line-by-line or it can be loaded (and automatically run) into the R workspace by using the source command:

```
source('scripts/zoop.depth.profile.r')
```

In addition to plotting, the figure will be saved in the subdirectory called `figures` in the rwork working directory.

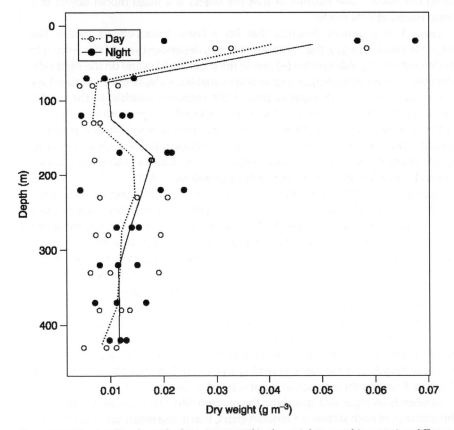

Figure 5.7 The density of zooplankton (measured in dry weight per cubic meter) at different depths and time of day, day and night. The lines show mean density at each depth, while symbols are the density of zooplankton in each sample.

5.5.4 Fish and Nekton

5.5.4.1 Hook-and-line

A data set containing information on catch of large fish predators is called *angling.stn.csv*. The data include information on station, date, position, time of day, fishing depth, time spent fishing, and catch of each species (g). We have made a script to plot catch per unit effort (CPUE, in kg) for fish captured by angling (Figure 5.8). The script can be opened and each line run. Alternatively, load the script and automatically run it with the source command:

```
source('scripts/angling.total.CPUE.r')
```

The script will import the data, plot it, and then save the plot to rwork/ figures. Note that this works when data are named exactly as in the example data, *angling.stn.csv*. The script can be modified to plot, for example, CPUE for one or a subset of species, CPUE at a particular station. To do this, open the script and modify the code accordingly.

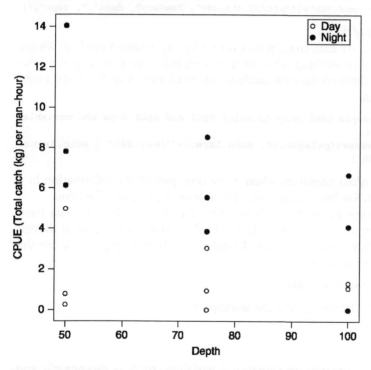

Figure 5.8 CPUE (kg per hour) for all fish species caught by hook-and-line.

5.5.4.2 Trawls

5.5.4.2.1 *Following Echo-layers*

The upper and lower echo layers (referred to as SSL1 and SSL2) may appear at different depths depending on time of day. We can formulate several questions or hypotheses to test with these data. We will use the data sets *PT.stn.csv*; which are the pelagic trawl data (station and catch data), and *PT.ind.csv*, the individual size measurements of various species sampled with the pelagic trawls. For each question, we will show how to load, explore, and analyze the data and interpret the results. As we begin to repeat methods, we will explain in less detail (to avoid repetition). Statistical methods used are explained only briefly and only to aid in interpretation of the data; further reading on the methods are encouraged and references are given.

Question 1 *Does the change from day to night affect the two SSLs differently in regard to their depth (i.e., does depth of SSL1 change between day and night compared to how the depth of SSL2 is changing?)?*

To test for a general (multispecies) diel vertical migration of organisms, we use a linear mixed-effects model (lme), which will account for the clustering of data within sampling stations (see Zuur *et al.*, 2009 for more details). The data used will be the station and catch data, not the individual size measurements.

1) *Import the data into R*

```
#Import all pelagic total catch samples (not individual measurements):
pelagic.df <- read.table('data/PT.stn.csv', header=T, dec='.', sep=',')
```

2) *Make a subset of the data*
   The `pelagic.df` data contain data from all pelagic station catches. We are only interested in working with the data sampled when following the echo layer because these contain the catches from SSL1 and SSL2. We make a subset of that data.

```
#Make a subsample that only contains SSL1 and SSL2 from the variable
  echo.layer:
SSLs.df <- subset(pelagic.df, echo.layer=='lower.SSL' | echo.layer=
  ='upper.SSL')
```

Subsetting is often necessary when using only part of the information in a larger data set. See Box 5.2 for general details on the `subset` function.

After using the `subset` function, a new data frame `SSLs.df` has been created. However, all factorial data retain their original levels even though some levels may contain no data. To get rid of these empty levels, use the `droplevels` function:

```
SSLs.df <- droplevels(SSLs.df)
```

Attach the data frame again to the workspace:

```
attach(SSLs.df)
```

3) *Plot the data*
   Using the plot function on categorical variables, such as *day.or.night* and *echo.layer*, will result in a boxplot. The boxplot displays the variation in *mean.depth* for each of these categories (Figure 5.9).

```
#Simple plot (left panel in Figure 5.9) for depth of SSLs depending
  on time of day
plot(mean.depth~interaction(day.or.night, echo.layer))
```

The syntax for plotting is given as y ~ x, where x is the interaction of the two variables *day.or.night* and *echo.layer*. Interactions can be plotted using the `interaction` function (see Box 5.3).

Box 5.2 Using the subset function

The `subset` function is very useful for extracting part of the data from a larger data set. Here, we make a smaller set of data `dataset2` from the larger data `dataset1`:

```
dataset2 <- subset(dataset1, Var1==3)
```

We pass two arguments to the `subset` function, the first is the dataset we want to subset (i.e. the larger `dataset1`), the second is the variable that contains the elements to be included (use a double equal sign) or excluded (specified by !=).

From the above example the original data set (`dataset1`) has a variable called `Var1`. We create a new object, called `dataset2`, which includes all rows of data where `Var1` has the value 3.

If `Var1` was a categorical variable, the value (or levels) to include must be enclosed in quotation marks :

```
dataset2 <- subset(dataset1, Var1=='placebo')
```

To include many variables, separate statements with the ampersand symbol (`&`). This only works when including information from different variables or when using the not equal syntax (!=):

```
dataset2 <- subset(dataset1, Var1=='placebo' & Var2=='G.morhua')
dataset2 <- subset(dataset1, Var1=='placebo' & Var2!='G.morhua')
```

This will not work if wanting multiple data (or levels) within the same variable, that is, `Var1=='placebo' & Var1=='real'`, because a subset cannot be equal to two values of one variable simultaneously. This would be the equivalent as saying: `1 = x = 2`.

Using values of one **or** another with the `subset` function is possible. In R, the 'or' symbol is the pipe symbol, the vertical bar: `|`. Now, instead of writing `Var1=='placebo' & Var1=='real'`, we can write `Var1=='placebo' | Var1=='real'`.

```
dataset2 <- subset(dataset1, Var1=='placebo' | Var1=='real')
```

Alternatively, use `%in%`:

```
dataset2 <- subset(dataset1, Var1 %in% c('placebo','real'))
```

Subsetting can become complicated quite quickly. If the subset includes multiple variables, or subsets a variable within a variable, use parentheses to keep terms for one variable together, similar to how mathematical formulae are expressed.

For example, we only want data where `Var1` is either `placebo` or `real`, but only from one sample site (`site A`); sample site is included in the variable `Var2`. To do this:

```
dataset2 <- subset(dataset1, (Var1=='placebo' | Var1=='real') &
    Var2=='siteA')
```

The smaller data set will now include data from only `site A`, which are coded as placebo or real. No other sites will be included nor will any other data in `Var1`.

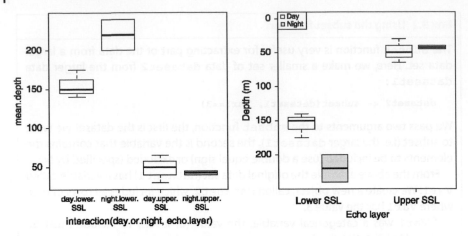

Figure 5.9 Median depth (thick black line) and depth range of echo-layers depending on time of day. The left panel shows the median depth (thick line), the upper and lower quartiles (upper and lower borders of the box), and the minimum and maximum values (upper and lower whiskers). Note that the surface of the water (0 m) is at the bottom of the y-axis. The right panel shows the same information in a slightly different way: 0 m is at the top of the y-axis, boxplots from day sampling are white and night are shaded grey.

The syntax makes a very simple plot and, confusingly, the surface of the water (0 m) is at the bottom of the y-axis. When plotting depth, it is more intuitive to have 0 m at the top of the y-axis. We can also make other adjustments to aid in the interpretation of the plot, such as color-coding the box-and-whisker plots to differentiate between day and night (Figure 5.9, right panel). See Box 5.3 for an explanation of the syntax for the right panel of Figure 5.9:

```
par(mar=c(5,5,2,2), las=1, cex.lab=1.5)
plot(mean.depth~interaction(day.or.night,echo.layer),
  ylim=rev(c(0,max(mean.depth))), xlab='Echo layer', ylab='Depth
  (m)', axes=F, col=c('transparent','gray80'), at=c(1,2,4,5))
axis(1, at=c(1.5, 4.5), labels=c('Lower SSL','Upper SSL'))
axis(2)
box()
legend(0.5,0, fill=c('transparent', 'grey80'), legend=c('Day','Night'))
```

Simple plots are always guaranteed to work. Data exploration should always begin with plotting the data. Simple plots will often give information that leads to more complex plots. Often defaulting to complex plots, including some of the plotting functions included here, can lead to errors if one does not fully understand one's data (or has not cleaned it of all errors).

4) *Data analysis*

Data analysis always begins with a question or hypothesis. Our question was:

Does the change from day to night affect the two SSLs differently with regard to their depth (i.e., does the depth of SSL1 change between day and night compared to how the depth of SSL2 changes)?

Box 5.3 Syntax of Figure 5.9 (right panel) explained

`par` is used to set graphical parameters. A full description of all the arguments that can be applied to this function can be found in the help (type `?par` in R). We used:

1) The graphical parameter `mar` to set the margins at the bottom, left, top, and right of the plot to 5, 5, 2, and 2 lines, respectively.
2) The function `c()` to combine information.
3) The graphical parameter `las` to change the orientation of axis labels. The options are: `las=0` parallel to the axis (default), `las=1` horizontal (as used in the example syntax for the right panel of Figure 5.9), `las=2` perpendicular to the axis, and `las=3` vertical.
4) The graphical parameter `cex.lab` to define the magnification of the x- and y-axis labels relative to the current setting of `cex`, which defaults to 1.

`plot` is a generic function for plotting that can be used to make many types of figures. As specified previously, the data to plot is written as y~x (or response ~ predictor).

1) The `interaction` function used within the `plot` function computes a factor that represents the interaction of the given factors and is especially useful when plotting the effect of two categorical predictors.
2) `ylim` is used to define y-axis limits, here from 0 to the maximum value of the variable called *mean.depth*.
3) The function `rev` reverses the axis so that zero is on top. This makes the plot more intuitive since zero depth starts at the top (water surface).
4) `xlab` and `ylab` are used to set names on the axes.
5) `axes=F` keep the axes from being plotted. We want to plot the axes manually (see below: `axis`).
6) `at` defines where each box will be drawn in the boxplot.

`axis` is used to draw axes after creating a plot with the axes suppressed (see above).

1) The first number within the `axis` function is used to define which axis to draw, where 1, 2, 3, and 4 refers to the lower, left, upper, and right axis.
2) The `at=c(...)` statement defines where R will draw tick marks on the axis (here at 1.5 and 4.5).
3) `labels=c(...)` defines the label for each tick mark (here, Lower SSL and Upper SSL). The reason to adjust the x-axis is to plot the two boxes for Lower SSL close to each other, followed by a larger space, and then the two boxes for Upper SSL.
4) `axis(2)` will plot a standard y-axis. No special modifications were needed.

`box()` includes a box around the plotting area. See the help on this function for special usage.

`legend` is used to place a figure legend on the plot.

1) The placement of the legend can by specified by giving values along the X- and then the y-axis, as done here `legend(0.5,0...)`. Instead of using

position, text can define placement, where the alternatives are: 'bottom-right', 'bottom', `bottomleft`, `left`, `topleft`, `top`, `topright`, 'right' and 'center'.

2) `fill` is used to define the colors for the filled symbols (default is the square).

3) legend specifies the text to use in the legend. Make sure that the text and color for the symbol (if used) are in the correct sequence (i.e., for this example, do not write c('Night' , 'Day') or a transparent symbol will plot for night and a dark symbol for day.

We assume the response variable, the depth data, are normally distributed with a constant variance. Depth is also a continuous variable (not categorical). This makes the model choice quite straightforward; use a model for a normal distribution and continuous response variable. We could perform separate tests for each echo layer; however, if we want to know if the change from day to night has a different effect on SSL1 compared to SSL2 (as in our question), we need to test for the interaction between the variables *day.or.night* and *echo.layer*. All predictors and the interaction are included in one model using the syntax:

```
#Model for depth of SSLs depending on time of day:
library(nlme)
fit2.lme <- lme(mean.depth~day.or.night*echo.layer, random=~+1|station)
```

To see a condensed version of the results, write:

```
anova(fit2.lme)
```

The output from the anova is:

	numDF	denDF	F-value	p-value
(Intercept)	1	3	222.1610	0.0007
day.or.night	1	3	3.6223	0.1531
echo.layer	1	3	402.5153	0.0003
day.or.night:echo.layer	1	3	26.5951	0.0141

The interaction is significant between the predictors *day.or.night* and *echo.layer* ($p = 0.0141$). A significant difference means that the effect of time of day depends on the echo layer (SSL1 or SSL2).

This information can be viewed in the summary output of R, which also includes parameter estimates with their associated standard errors, statistics (for this test, it is the t-statistic value), and p-values. Use the `summary` function as follows:

```
summary(fit2.lme)
```

Output:

```
Random effects:
 Formula: ~+1 | station
        (Intercept) Residual

StdDev:    15.27525 10.63929
```

```
Fixed effects: mean.depth ~ day.or.night * echo.layer
                                        Value Std.Error DF   t-value p-value
(Intercept)                         155.00000 10.747523  3 14.421927 0.0007
day.or.nightnight                    65.00000 16.993327  3  3.825031 0.0315
echo.layerupper.SSL                -106.66667  8.686942  3 -12.278966 0.0012
day.or.nightnight:echo.layerupper.SSL -70.83333 13.735261  3 -5.157043 0.0141
```

An explanation of this output is in Box 5.4.

To fully understand the summary output, think about it in terms of the functional expression for the model:

$$depth_{ij} = \beta0 + \beta1 day.or.night_i + \beta2 echo.layer_j + \beta3 day.or.night_i:echo.layer_j + \varepsilon_{ij}$$

where each beta (β) is an estimate in the model output.

```
                                       Estimate
(Intercept)                            155.00000
day.or.nightnight                       65.00000
echo.layerupper.SSL                   -106.66667
day.or.nightnight:echo.layerupper.SSL  -70.83333
```

This information is then used to estimate the depth for each SSL during the day and during the night. It's best explained by working through an example:

What is the depth at day for the upper SSL?

To estimate this depth, we first include the $\beta0$ value because this is the reference point (the mean of the group that is first in the alphabet when predictors are categorical).

```
155.0000 +…
```

The next estimate refers to the change in mean when going from day to night, but we want the value for day, therefore we multiple the estimate for $\beta1$ by 0 before adding it to the $\beta0$ estimate:

```
155.0000 + (65.00000*0) +…
```

The next estimate describes the change in the mean when going from the lower to the upper SSL during day. This is exactly what we want, so we multiple the estimate for $\beta2$ by 1 before adding it to the $\beta0 + \beta1$ estimate:

```
155.0000 + (65.00000*0) + (-106.66667*1) +…
```

Finally, we need to add the information from the last term, $\beta3$, was the difference between the change in depth of the lower SSL going from day to night and the change in the depth of the upper SSL from day and night. But this term includes information about night and we are only interested in day, so we multiple $\beta3$ by 0 before adding it.

```
155.0000 + (65.00000*0) + (-106.66667*1) + (-70.83333*0)
```

This gives a result of 48.33333. This can be roughly checked by looking at the boxplot, but be aware that the horizontal line in the plot represents the median and we estimated the mean (Figure 5.9).

In addition to the interaction term (i.e., the difference in the differences of the change), we are interested in the depth difference between day and night within

---

Box 5.4 Interpretation of the summary output for lme

`Random effects`: accounts for pseudoreplication or the effects of correlated samples within our model. Station was included as a random effect factor (the random effect of station was nested within the model). By doing this, we have tried to account for samples within stations having a higher correlation (being more alike) than samples taken at different stations.

`StdDev`: gives the variability caused by stations compared to the unexplained variability of the model (residual). In this example, variability due to stations (standard deviation = 15.27525) is larger than the residual standard deviation (10.63929). The residual variance is giving information about the amount of variability within the variables `day.or.night` and `echo.layer`. The variance for the random effect of station gives information on how much of the within `day.or.night` and `echo.layer` variance is explained by station.

`Fixed effects`: these are the effects associated with the predictors in the model.

`Intercept`: the intercept is the mean of the group that is first in the alphabet when predictors are categorical. Here, that is `echo.layer=lower.SSL` and `day.or.night=day`. Thus, the estimated intercept (155.00000) is simply the mean depth for the lower SSL at daytime. The associated p-value (0.0007) describes if the estimated value (155.00000) is different from zero or not. This p-value does not contain useful information in this particular example.

`day.or.nightnight`: describes the change in the mean when going from day to night in the lower SSL. Thus, the lower SSL is 65 meters deeper than at daytime. This difference is statistically significant (p = 0.0315).

`echo.layerupper.SSL`: describes the change in the mean when going from the lower to the upper SSL at daytime. This change is significant (p = 0.0012).

`day.or.nightnight:echo.layerupper.SSL`: describes the difference in the change when going from lower SSL at day to lower SSL at night compared to going from upper SSL at day to upper SSL at night [maybe better visualized as (the change in going from lower SSL at day to lower SSL at night) – (going from upper SSL at day to upper SSL at night)]. The associated p-value states that the difference in the change is significant (p = 0.0141). Because there are only two levels for each of the two predictors, this p-value is the same as p-value for the interaction from the anova shown previously.

---

each echo-layer. Our model results indicated that the lower SSL had a statistically significant difference in mean depth between day and night (p = 0.0315). Because a normal summary output never shows more than the maximum orthogonal contrasts, we need to change the contrast matrix to get the p-value for the difference between day and night for the upper SSL. The easiest way to do this is to change the order of the levels of the variable *echo.layer* and rerun the analysis. Reordering a categorical variable is done using the `relevel` function, where the level we want ordered first is set using the `ref` argument:

```
# reorder the categorical variable echo.layer
echo.layer2 <- relevel(echo.layer, ref='upper.SSL')
# rerun the model, giving it a different name
```

```
fit2b.lme <- lme(mean.depth~day.or.night*echo.layer2, random=~+1|
  station)
# view the output
summary(fit2b.lme)
```

Looking at only the fixed effects output from the summary command, the estimated depth difference between day and night for the upper SSL (highlighted in orange) is 5.83333 m shallower at night than at daytime. This difference is not statistically significant (p = 0.7540).

```
Fixed effects: mean.depth ~ day.or.night * echo.layer2
                                       Value Std.Error DF  t-value p-value
(Intercept)                         48.33333 10.747523  3  4.497160  0.0205
day.or.nightnight                   -5.83333 16.993327  3 -0.343272  0.7540
echo.layer2lower.SSL               106.66667  8.686942  3 12.278966  0.0012
day.or.nightnight:echo.layer2lower.SSL 70.83333 13.735261  3  5.157043  0.0141
```

Conclusion: Organisms in the lower SSL are found 65 m deeper at night than during the daytime, the depth difference between day and night is less for organisms in the upper SSL. In the upper SSL, the response to time of day is absent, that is, organisms are only 5.8 m higher in the water column at night compared to day. This depth difference in the upper SSL is not statistically significant (p = 0.7540).

Question 2 *Does the size distribution of pearlside change with SSL or time of day?* We are interested in investigating the size distribution of *M. muelleri* within each of the echo-layers and seeing whether this size distribution changes with time of day. For this analysis, we need to import the individual size measurements, found in the *PT.ind.csv* file. We can phrase the question as the following null hypothesis:

> *H0: There is no change in the size of* M. muelleri *regardless of which echo-layer it inhabits and/or the time of day.*

First, we import the data:

```
#Import data for individual size measurements
pelagic.ind.df <- read.table('data/PT.ind.csv', header=T, dec='.', sep=',')
```

Next, make a data set consisting only of data from following the echo-layer. We subset our data:

```
#Make a subset that only contains SSL1 and SSL2:
SSLs.ind.df<-subset(pelagic.ind.df,echo.layer=='lower.SSL'|echo.layer=='upper.SSL')
SSLs.ind.df <- droplevels(SSLs.ind.df)
attach(SSLs.ind.df)
```

But remember, we are only interested in the pearlside, *M. muelleri*. We therefore make another data set. Here, we are showing it as a separate step, but one could subset on both *echo.layer* and *species* in the same command line.

```
#Make a subset that only contains M. muelleri:
#Note! get variable names by: names(SSLs.ind.df)
#get species names by: levels(species)
SSL.M.muelleri.df <- subset(SSLs.ind.df, species=='M.muelleri')
attach(SSL.M.muelleri.df)
```

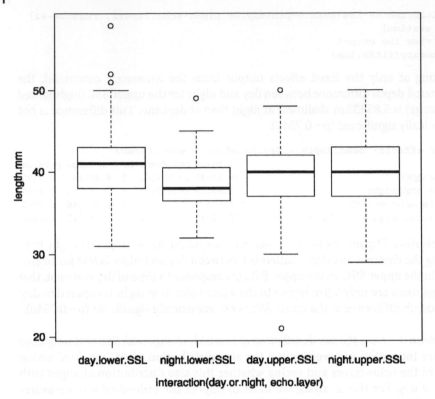

Figure 5.10 Size of *M. muelleri* caught in the upper and lower echo-layers during the day and night.

Make a simple plot showing the range of length values for pearlside from each SSL at day and at night as in Figure 5.10:

```
#Simple plot:
plot(length.mm~interaction(day.or.night,echo.layer))
```

A boxplot showing the depth range (not length) of pearlside in each of the SSLs for day and night can also be produced, following similar syntax as that used to make Figure 5.9. Such a figure could be useful to interpret the analysis or to decide which SSL to target to catch a large amounts of pearlside for sampling.

Next, load the nlme library, if not already loaded in R. Because the example data have been checked for errors, we proceed directly to the analysis. Create the model object and produce a condensed view of the results using anova. The model formulation is similar to the previous analysis.

```
# load the library, if not already loaded
library(nlme)
#Analysis of size distribution of echo layers depending on day or night:
fit2.lme <- lme(length.mm~day.or.night*echo.layer, random=~+1|
  station)
anova(fit2.lme)
```

Output:

```
                     numDF denDF  F-value p-value
     (Intercept)         1   667 5174.512 <.0001
     day.or.night        1     3    0.160  0.7159
     echo.layer          1   667   12.476  0.0004
     day.or.night:echo.layer 1 667  27.170 <.0001
```

The interaction term is statistically significant (p < 0.0001). This means that the change in size of *M. muelleri* between day and night is different for the two echo-layers.

Now look at the parameter estimates of the model and compare with the plot. Use the summary output:

```
summary(fit2.lme)
```

Output:

```
Random effects:
Formula: ~+1 | station
        (Intercept) Residual
StdDev:   1.200231 3.879263

Fixed effects: length.mm ~ day.or.night * echo.layer
                                    Value Std.Error  DF t-value p-value
(Intercept)                      41.13844 0.7326995 667 56.14641   0.000
day.or.nightnight                -2.99963 1.3157752   3 -2.27975   0.107
echo.layerupper.SSL              -2.47250 0.4349078 667 -5.68511   0.000
day.or.nightnight:echo.layerupper.SSL 4.47725 0.8589426 667 5.21252   0.000
```

We can see some of the parameters are statistically significant, but are the differences *biologically significant*? To make a decision about this, use information from the parameter estimates and the created figures.

Question 3 *Does the biomass of organisms captured in the SSLs depend on the layer type or time of day?*
To fully answer the question, we need to return to the station data, which contains the total catch of each species. If this data is no longer in the R workspace (i.e., if R was closed without saving the workspace between this and the previous analysis) re-import the data. Otherwise, attach the data.

```
attach(SSLs.df)
```

Adjust for effort (time spent trawling) before comparing biomass. The number or weight caught during, for example, 15-minutes of trawling will be different from, for example, 30-minutes; results need to be standardized to the same scale. Here, we standardize all catches to 1-hour of trawling:

```
#Create a variable describing effort (in hours):
SSLs.df$hours.trawling<-difftime(strptime(endtime.local,"%H:%M"),
strptime(starttime.local,"%H:%M"), units='hours')
attach(SSLs.df)
```

The function difftime estimates the difference between *endtime.local* and *starttime.local* in hours (units='hours'); strptime converts the time variables to time in hours and minutes, separated by a colon (:). Reattach the data to make sure the changes are in the workspace and available for the analysis.

```
#Fix errors in time estimates when sampling crosses over midnight
  e.g. starts at 23:57 one day and ends at 00:07 the next date
#(The programming is far from elegant, but it works). Here, the
  syntax has been separated into different lines and indented to
  aid readability.
SSLs.df$hours.trawling <- ifelse(hours.trawling<0,
  difftime(strptime('23:59',format="%H:%M"),strptime(starttime.local,
    format="%H:%M"),units='hours')
  +
  difftime(strptime(endtime.local,"%H:%M"), strptime('00:01',
  format="%H:%M"), units='hours')
  +
  0.03333333 , hours.trawling)
attach(SSLs.df)
```

The `ifelse` function was used, which works by: *if* expression is true, do this, *else* do that. Every occurrence where *hours.trawling* is less than 0 is replaced with a correction that accounts for crossing midnight. If the *if* statement is not true, the *else* statement states to use the originally estimated *hours.trawling* (i.e., leave the value as is).

```
#Check no negative values and no extremely high or low values exist:
hours.trawling
# for larger data sets, where it is not practical to view all values,
  use one of the alternates below
summary(hours.trawling)
table(hours.trawling)
```

Create a new variable that is catch per hour trawling by dividing catch (in grams) by the hours trawling, then convert to kg by dividing by 1000.

```
#Create a variable that is catch per hour in kg.
#Divide by 1000 to convert the unit from g to kg:
biomass.per.hour <- (totalcatch.g/as.numeric(hours.trawling))/1000

#Plot of catch within SSLs depending on time of day:
plot(biomass.per.hour~interaction(day.or.night, echo.layer),
ylab='Catch (kg per hour trawling)', main='Total catch')
```

The resulting plot is shown in Figure 5.11.
Use the `tapply` function (see Box 5.5) to estimate the number of trawl tows taken on each of the SSL layers at each time of day.

```
#Estimate sample size (number of independent samples) for each group:
tapply(!is.na(biomass.per.hour), list(day.or.night, echo.layer), sum)
```

For the analysis, we again use a linear mixed-effects model to account for the clustering of data within sampling stations:

```
library(nlme)
fit3.lme <- lme(biomass.per.hour~day.or.night*echo.layer,
  random=~+1|station,na.action='na.omit')
anova(fit3.lme)
```

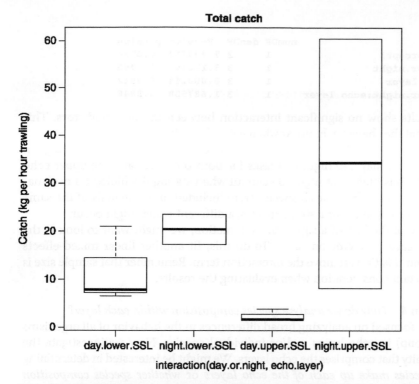

Figure 5.11 Total biomass as catch per hour (kg) in each echo-layer, lower and upper, during day and night. The sample size for each group, left to right, is 3, 2, 3, and 2 hauls for each of the layers.

Box 5.5 `tapply`

`tapply` applies a function to each group of values given by a unique combination of the levels of certain factors. In the example above, we estimate the length (n) of the variable *biomass.per.hour* given the levels in the variables *day.or.night* and *echo.layer*, that is, sample size within the four groups in Figure 5.11.

According to the help syntax for `tapply` (write `?tapply`), a more logical syntax than what was used would be the following:

```
tapply(biomass.per.hour, list(day.or.night, echo.layer), length,
    na.rm=T)
```

This uses the `length` function within `tapply` and specifies that the function should not count missing values (`na.rm=T`). But unlike many other functions, e.g. `sum`, `mean`, the `length` function does not allow `na.rm` as an option.

The syntax using the function `sum` is a workaround for this. Logical values `TRUE` and `FALSE` are obtained when writing `!is.na(biomass.per.hour)` and are then counted as 1 and 0 within the `sum` function.

Output:

	numDF	denDF	F-value	p-value
(Intercept)	1	3	7.322553	0.0734
day.or.night	1	3	3.120640	0.1755
echo.layer	1	3	0.000144	0.9912
day.or.night:echo.layer	1	3	1.687508	0.2848

The results show no significant interaction between the two predictors. This means that the change in biomass when going from day to night does not depend on the echo layer (either SSL1 or SSL2). As seen in Figure 5.11, the change in catch between day and night increases for both the lower and the upper echo layers; catch is highest at night. Be careful when stating the biological explanation for this result because all species were included in the analysis. If the same analyses is completed for a single species, a different result might occur.

Because of the lack of a significant interaction, we might want to look at the effect of each predictor on catch. To do this, fit another linear mixed-effects model, but this time, remove the interaction term. Remember that sample size is an important consideration when evaluating the results.

Question 4   *How do we analyze species composition within each layer?*
We have focused on analyzing broad differences in the behavior of all organisms (as a group) in echo layers and/or individual species. Now we investigate the community that comprises the echo layers. We might be interested in determining *what species make up each of the echo layers* or *whether species composition differs in the upper echo layers compared to the lower layers.*

The first step to answering these questions is to plot the data. As specified previously, before beginning any analysis, plot the data. We have made a script that makes several figures. Open the script to run it line by line or run it directly (in its entirety) by using the source command:

```
source('scripts/species.composition.SSLs.r')
```

In addition to plotting, the figures will automatically be saved in the rwork / figures subdirectory.

The script, similar to the previous analyses, reads in the data and creates a few variables. Plots of the proportion total biomass (Figure 5.12) and numbers (Figure 5.13) of each species in each layer for day and night trawling. These scripts plot the raw data; ideally, one should standardize to a common unit of trawling effort (e.g., 30-min, 60-min) or (where possible) estimate abundance per unit of swept area.

5.5.4.2.2   *Species Pelagic Depth Distribution*
The pelagic trawl data contain information from many different types of sampling; now we analyze samples taken from district depth ranges. The data are grouped into depth bins or fixed depth ranges and samples have been standardized to a common unit of effort (e.g. 1-h trawling).

Overview of Total Catch of a Species at Different Depths and Time of Day
A script *PT.catch.of.all.species.r* has been written to plot biomass (or numbers) per hour trawling for each species captured either during day or

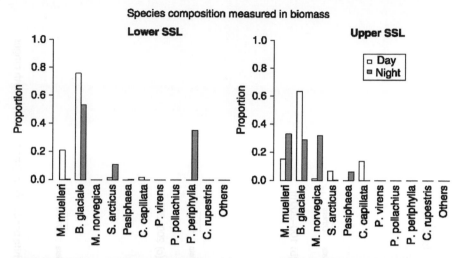

Figure 5.12 The proportion of total biomass of each species caught during day and night in the upper and lower echo layers. Data have not been standardized to a similar unit of effort before plotting.

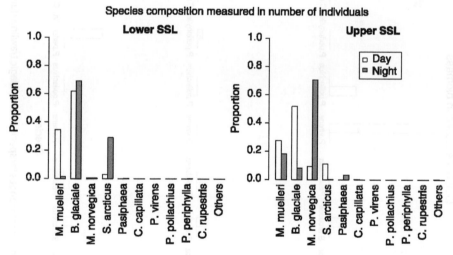

Figure 5.13 The proportion (in numbers) of each species caught during day and night in the upper and lower echo layers. Data have not been standardized to a similar unit of effort before plotting.

night at three depth ranges: 0–100 m, 100–200 m, and 200–300 m. The data for the fixed depth ranges are identified in the variable *sampling.type*. As with previous examples, the script can be opened in an editor, where each command can be run separately, or run in its entirety by using source:

```
source('scripts/PT.catch.of.all.species.r')
```

Catch per hour trawling in biomass (Figure 5.14) and numbers (Figure 5.15) show a similar picture. Pearlside (*M. muelleri*) were captured in high amounts (both in

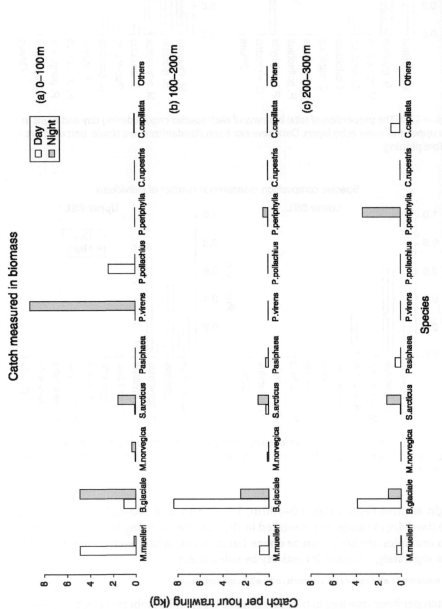

Figure 5.14 Catch (kg) per hour at three depth ranges by species from pelagic trawling during the day (white bars) or night (shaded bars). Depth ranges were 0–100 m, 100–200 m, and 200–300 m.

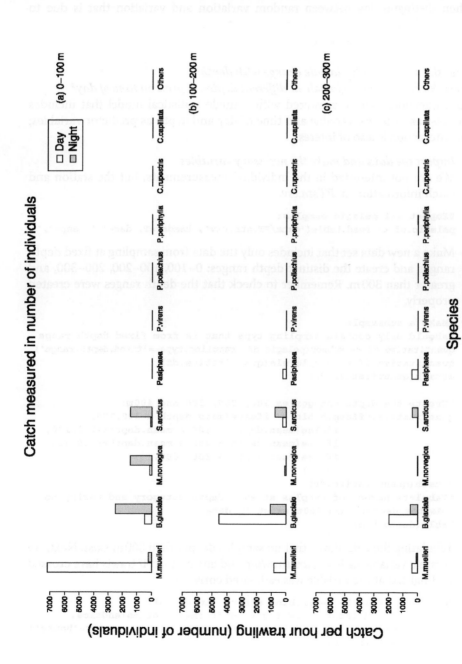

Figure 5.15 Catch (numbers) per hour at three depth ranges by species from pelagic trawling during the day (white bars) or night (shaded bars). Depth ranges were 0–100 m, 100–200 m, and 200–300 m.

kg and numbers) in surface waters during the day; they were not captured at night, likely because they were very close to the surface where capture is difficult. Other mesopelagic organisms were captured in high amounts in surface waters at night, but deeper depths during day. Some species dominate more than others and warrant further investigation; one of these is the lanternfish *Benthosema glaciale*.

The figures do not give any information about the variation among independent samples from a particular depth or time of day. This is important information when distinguishing between random variation and variation that is due to statistically significant differences between the groups.

This then leads to us to ask:

Does the biomass of B. glaciale change with depth?
Does the biomass of B. glaciale at different depths depend on time of day?
Both questions can be answered with a single statistical model that includes biomass as response variable and time of day and depth as predictor variables; the interaction is also of interest.

1) *Import the data and make the necessary variables*
   We are not interested in the individual measurements, but the station and catch information in *PT.stn.csv*.

```
#Import all pelagic samples:
pelagic.df <- read.table('data/PT.stn.csv', header=T, dec='.', sep=',')
```

Make a new data set that includes only the data from sampling at fixed depth ranges and create the distinct depth ranges: 0–100, 100–200, 200–300, and greater than 300 m. Remember to check that the depth ranges were created properly.

```
#Make a subsample
#should only contain sampling type that is from fixed depth range
quantitative.df <- subset(pelagic.df, sampling.type=='fixed.depth.range')
quantitative.df <- droplevels(quantitative.df)
attach(quantitative.df)

#Create the depth categories 100, 200, 300 and 400m:
quantitative.df$depth.bin <- ifelse(mean.depth <=100,100,
                ifelse(mean.depth > 100 & mean.depth<=200,200,
                ifelse(mean.depth > 200 & mean.depth<=300,300,
                ifelse(mean.depth > 300, 400,
                NA))))
attach(quantitative.df)
#Tabulate number of samples at each depth category and verify no
  depths deeper than 400m exist in data:
table(depth.bin)
```

Tabulating the data shows that no samples deeper than 300 m exist. Next, we create a variable called *hours.trawling* and ensure that, if trawls have crossed midnight, that the variable was estimated correctly.

```
#Create a variable describing effort (in hours):
quantitative.df$hours.trawling <- difftime(strptime(endtime.
  local,"%H:%M"), strptime(starttime.local,"%H:%M"), units='hours')
attach(quantitative.df)
```

```
#Fix errors in time estimates when sampling goes over two dates,
  e.g. #starts at 23:57 one date and ends at 00:07 the next date
quantitative.df$hours.trawling <- ifelse(hours.trawling<0,
  difftime(strptime('23:59',format="%H:%M"), strptime(starttime.
  local,format="%H:%M"),units='hours')+ difftime(strptime(endtime.
  local,"%H:%M"),strptime('00:01',format="%H:%M"), units='hours')
  + 0.03333333, hours.trawling)
attach(quantitative.df)

#Check that no negative values exist and that all values appear
  within the expected range:
summary(hours.trawling)
```

Standardize the catch to 1-hour trawling and convert catches from grams to kg.

```
#Create a variable that measures catch per hour in kg.
#Divide by 1000 to convert the unit from g to kg.:
quantitative.df$Kg.B.glaciale.per.hour<-(B.glaciale.g/as.numeric(hours.
  trawling))/1000
attach(quantitative.df)
```

2) *Plot the data*

Plot the data and estimate the sample size for each group. This is one way to quickly error check the data and visualize the amount of variation between the sample groups.

```
#Simple plot: overview of the variation within each group of samples:
plot(Kg.B.glaciale.per.hour~interaction(day.or.night, depth.bin))
#tabulate number of samples within each group:
tapply(!is.na(Kg.B.glaciale.per.hour), list(day.or.night, depth.bin), sum)
```

A slightly more polished version of the plot can be made using the script *PT.catch.of.individual.species.r* (Figure 5.16). Open the script and run each command or source the entire script into R:

```
source('scripts/PT.catch.of.individual.species.r')
```

The script was written specifically for *B. glaciale*. To make plots for any other species, the script must be modified.

3) *Analyze the data*

The question we want to answer is: *Does the biomass of* B. glaciale *change with depth and time of day?* As with previous analyses, we must account for the clustering of data within sampling stations; therefore, we use a linear mixed-effects model with station as a random effect. Only the data for biomass are used here. To run the analysis requires more than one sample per group, where group is defined by depth range and time of day (see Figure 5.16).

```
#load the library to access the lme function
library(nlme)
fit5.lme <- lme(Kg.B.glaciale.per.hour~day.or.night*as.
  factor(depth.bin), random=~+1|station, na.action='na.omit')
anova(fit5.lme)
```

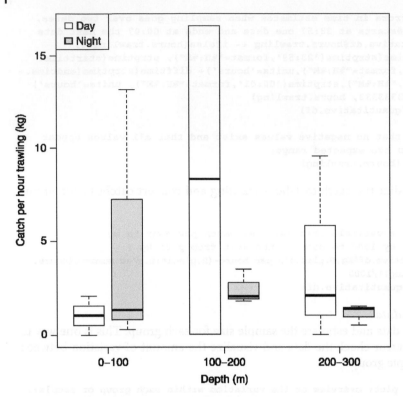

Figure 5.16 Catch per hour of *B. glaciale* (biomass) from day (white) and night (shaded) for three depth ranges: 0–100 m, 100–200 m, and 200–300 m. Solid black lines represent median values.

Output:

```
                                  numDF denDF  F-value p-value
(Intercept)                         1     7   4.965010  0.0611
day.or.night                        1     4   0.105467  0.7616
as.factor(depth.bin)                2     7   0.348624  0.7172
day.or.night:as.factor(depth.bin)   2     7   1.562849  0.2747
```

None of the predictors were significant. This is not surprising; the data for each group show a large range of variability, often overlapping (Figure 5.16). The number of independent samples (groups) is low, which may explain the non-significant result.

The same procedure can be followed for other species in the samples.

Question 2 *Does the size distribution of* B. glaciale *change within the district depth ranges, with time of day?*
After investigating differences in catch with depth range and time of day, the next step is to determine whether there may be differences in the size distribution of organisms related to depth range or time of day. The individual size measurement, found in the *PT.ind.csv* data file, are needed for this analysis. We can phrase the question as the following null hypothesis:

H0: The size of B. glaciale *at different depths does not depend on time of day (day or night).*

First, we import the data:

```
#Import data of individual size measurements
pelagic.ind.df <- read.table('data/PT.ind.csv', header=T, dec='.', sep=',')
```

Make a data set of only the data needed:

```
#Make a subsample - contains data from fixed depth range
quant.ind.df <- subset(pelagic.ind.df, sampling.type=='fixed.depth.range')
quant.ind.df <- droplevels(quant.ind.df)
attach(quant.ind.df)
```

Create the depth categories:

```
#Create the depth categories 100, 200, 300 and 400m:
quant.ind.df$depth.bin <- ifelse(mean.depth <=100,100,
                 ifelse(mean.depth > 100 & mean.depth<=200,200,
                 ifelse(mean.depth > 200 & mean.depth<=300,300,
                 ifelse(mean.depth > 300, 400,
                 NA))))
attach(quant.ind.df)


#Tabulate numbers of samples at each depth category and verify no
  samples from depths deeper than 400m:
table(depth.bin)
```

Make another subset, selecting the species of interest:

```
Bglac.df <- subset(quant.ind.df, species=='B.glaciale')
Bglac.df <- droplevels(Bglac.df)
attach(Bglac.df)
```

Plot the data:

```
#Simple plot:
plot(length.mm~interaction(day.or.night,depth.bin))
```

A slightly more polished version of the plot can be made by using the syntax made to plot Figure 5.9 and in Box 5.3 or by using the script *PT.size.of.individual.species.r* (Figure 5.17). Open the script and run each command or source the entire script into R:

```
source('scripts/PT.size.of.individual.species.r')
```

The script was written specifically for *B. glaciale*. To make plots for any other species, the script must be modified.

Before performing the statistical test, check the sample size within each group:

```
tapply(!is.na(length.mm), list(day.or.night, depth.bin), sum)
```

The output of this indicates there is adequate data in each group to proceed with the analysis.

```
        100 200 300
day     174 241 240
night   274 358 253
```

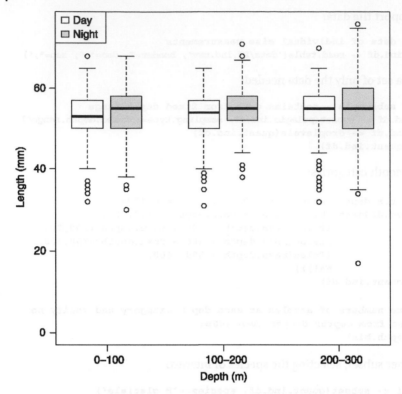

Figure 5.17 Length (mm) of *B. glaciale* from day (white) and night (shaded) for three depth ranges: 0–100 m, 100–200 m, and 200–300 m. Solid lines indicate median values.

Use a linear mixed-effects model, including station as a random effect, to test if the size of *B. glaciale* depends on discrete depth range and time of day:

```
library(nlme) #to access the lme function
fit6.lme <- lme(length.mm~day.or.night*as.factor(depth.bin),
  random=~+1|station)
anova(fit6.lme)
```

Output:

	numDF	denDF	F-value	p-value
(Intercept)	1	1530	6528.092	<.0001
day.or.night	1	4	0.759	0.4329
as.factor(depth.bin)	2	1530	2.242	0.1066
day.or.night:as.factor(depth.bin)	2	1530	5.008	0.0068

A significant interaction between time of day and depth range exists (p = 0.0068). This means that the change in size between depth range is statistically different between day and night samples.

Take a closer look at the parameter estimates of the model and the size of the effects using the summary function on the model object:

```
summary(fit6.lme)
```

Output:

```
Linear mixed-effects model fit by REML
 Data: NULL
       AIC      BIC     logLik
  10189.32 10232 -5086.659

Random effects:
 Formula: ~+1 | station
          (Intercept) Residual
StdDev:    1.562892   6.57315

Fixed effects: length.mm ~ day.or.night * as.factor(depth.bin)
```

	Value	Std.Error	DF	t-value	p-value
(Intercept)	53.64514	1.0653671	1530	50.35367	0.0000
day.or.nightnight	-0.58954	1.4577862	4	-0.40441	0.7066
as.factor(depth.bin)200	-1.24528	0.8388098	1530	-1.48458	0.1379
as.factor(depth.bin)300	-0.86781	0.8627237	1530	-1.00590	0.3146
day.or.nightnight:as.factor(depth.bin)200	3.12735	0.9963474	1530	3.13882	0.0017
day.or.nightnight:as.factor(depth.bin)300	1.95514	1.0535306	1530	1.85580	0.0637

See Box 5.4 for details on how to interpret output from linear mixed-effects models. Plotting the data will aid in the interpretation.

5.5.4.2.3 Diet of Predators

Understanding the feeding ecology of a species gives insight into its population dynamics, but also leads to better understanding of resource partitioning, inter- and intraspecies dynamics, habitat preferences, and energy transfer within and between ecosystems (Braga, Bornatowski, and Vitule, 2012). The work involved in analyzing gut contents is necessary to get a deeper insight to population and ecosystem dynamics.

The most common predators in Masfjord are saithe (*Pollachius virens*) and pollack (*Pollachius pollachius*) and we are interested in determining if mesopelagic organisms make up a dominant part of their diet. The data, *diet.predators.csv*, contains data on predator species (length, weight, sex, liver weight), prey items, state of digestion of prey, and whether the predators were feeding, had empty stomachs, or had regurgitated the contents (called *stomach.status*).

1) *Import the data*

```
#Import the main data set:
diet.df <- read.table('data/diet.predators.csv', header=T, sep=',')
```

2) *Create a smaller data set of only saithe*
   Create a data set of only five prey species from saithe (*Pollachius virens*), omitting information from individuals that have regurgitated their contents; prey are given as total weight in g:

```
# The syntax has been separated into 3 lines to aid readability
P.virens.gut.content.g.df <- subset(diet.df,
species=='P.virens' & stomach.status!='regurgitated',
select=c(M.muelleri.g, M.norvegicus.g, shrimp.g, B.glaciale.g,
  other.fish.g))
```

Attach the data:

```
attach(P.virens.gut.content.g.df)
```

Create a simple plot using the `boxplot` function:

```
boxplot(P.virens.gut.content.g.df)
```

The script *diet.predators.r* contains additional syntax to improve the plot. Alternatively, the plot can be created and saved by:

```
source('scripts/diet.predators.r')
```

The script will create boxplots of prey items in grams and numbers for saithe and pollack (Figure 5.18). This information can be used to broadly note which prey items (of those chosen for plotting) are important for these two predators. Before making any conclusions about differences in diet between the two species, the sample size of the predator must be considered.

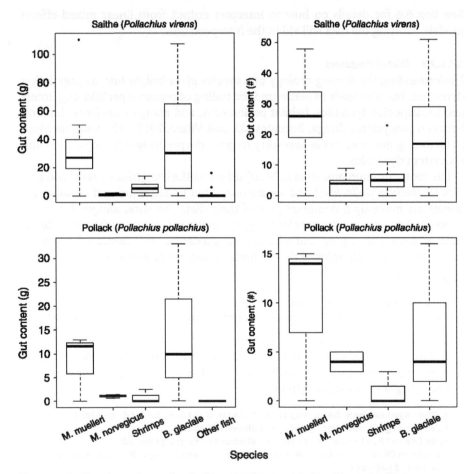

Figure 5.18 Abundance (g, numbers) of five prey items eaten by saithe and pollack.

5.5.4.2.4 *Bottom Trawls*

The number of bottom trawls were few because of the need to minimize the disturbance of bottom living species. Because the amount of data is sparse, we are restricted to simple summaries and plots to describe catch in biomass, number of individuals, and size of individuals.

A script *BT.catch.of.all.species.r* has been provided to plot catch in biomass and numbers. The script can be opened and run line-by-line or sourced into the workspace:

```
source('scripts/BT.catch.of.all.species.r')
```

Catch has been standardized to per hour trawling and weight has been converted to kg. The catch is made up of many species, but four deep-water species and the jellyfish, *P. periphylla*, dominate the catch in weight (Figure 5.19). The catch in numbers gives a slightly different picture of the trawl catches; the myctophid, *B. glaciale*, dominates the catches (Figure 5.20). Myctophids are small fish and while there were many, their combined weight was less than some of the larger species, such as the roundnose grenadier, *C. rupestris*, or the velvet belly lanternshark, *E. spinax*.

Objects containing the estimated number of hauls (i.e., the number of independent samples) for each species was estimated in the *BT.catch.of.all.species.r* script. The objects can be viewed by typing:

```
BT.samplesize.biomass
BT.samplesize.counts
```

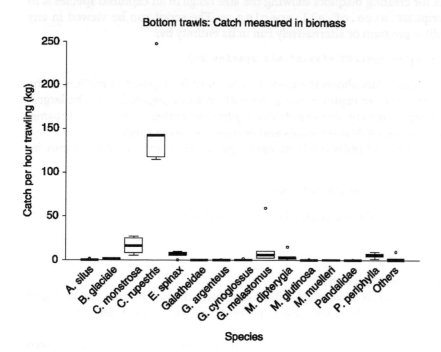

Figure 5.19 Catch (kg per hour of trawling) of species captured by bottom trawls.

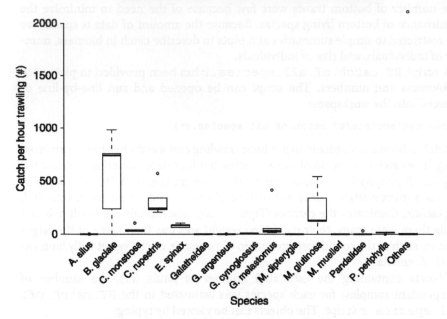

Figure 5.20 Catch (number of individuals per hour of trawling) of species captured by bottom trawls.

Syntax for creating boxplots showing the size range of all captured species is in the script `BT.size.of.all.species.r`. The script can be viewed in any text editor program or alternatively run in its entirety by:

```
source('scripts/BT.size.of.all.species.r')
```

The resulting figure shows the median, interquartile ranges, and outliers in the size of each species captured using the bottom trawl (Figure 5.21). The largest species captured were blue ling, *Molva dypterygia*, saithe, *P. virens*, and brosme, *Brosme brosme*, while myctophids and pearlside were the smallest.

The number of individuals of each species that were measured can be found by:

```
tapply(!is.na(length.mm), species, sum)
```

This assumes the data were attached to the workspace.

5.5.5 Mapping

5.5.5.1 Making Maps with R

Many ways exist to make maps with R and much documentation can be found by searching the Internet. Many different packages exist and it can be overwhelming.

This section does not focus on spatial statistics or the making of complicated maps, but rather how to make simple maps of a study area and mark sampling

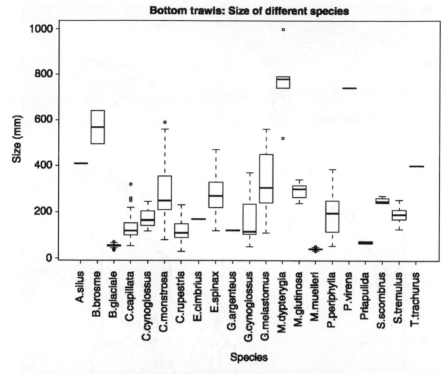

Figure 5.21 Boxplots showing the size range (mm) of all species caught using bottom trawls.

stations. The R libraries needed for the following examples include `mapdata` (Becker, Wilks, and Brownrigg, 2016a), `maps` (Becker *et al.*, 2016b), `sp` (Pebesma and Bivand, 2005; Bivand, Pebesma, and Gomez-Rubio, 2013) and `RgoogleMaps` (Loecher and Ropkins, 2015).

For some of the examples, the maps created are based on maps from the following external sources: Global Administrative Areas (GADM; http://www. gadm.org) and Google Maps. Both GADM and Google Maps are liberal with the permissions for using their maps in academic publications. R can use other external map sources, for example, shape files from The Norwegian Mapping Authority at http://www.kartverket.no/en/. Always check guidelines and permissions for using external map sources.

Example: Plotting stations from an oceanic survey.
We want to create a map of our sampling locations while on survey in the North Atlantic and Norwegian Sea. Load the libraries called `maps` and `mapdata`. If these are not yet installed, refer to Section 5.5.1. The syntax below will create a map of the North Atlantic and Norwegian Sea; trawl stations are marked as red dots (Figure 5.22).

```
#################
# North Atlantic #
#################
```

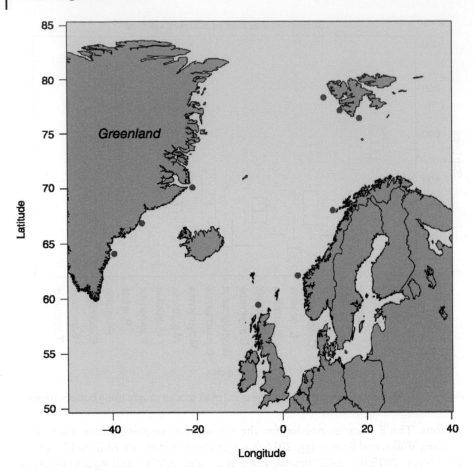

Figure 5.22 A map of the North Atlantic showing bottom trawling locations as red dots.

```
#Load required libraries:
library(maps)
library(mapdata)

#The following three lines color the sea blue
#xlim and ylim set the longitude and latitude limits:
map('worldHires', xlim=c(-50,40), ylim=c(50,85), resolution=0, type='n')
u <- par('usr')
rect(u[1], u[3], u[2], u[4], col='lightblue')

#Add in land masses:
map('worldHires', xlim=c(-50,40), ylim=c(50,85), fill=T, col='gray60',
   resolution=0, add=T)

#Add axes showing longitude and latitude values:
map.axes()

#Add titles to the axes:
mtext('Longitude', side=1, line=2.2, cex=2)
mtext('Latitude', side=2, line=2.2, cex=2, las=0)
```

```
#Add text: add name `Greenland' to the map:
text(-39, 75, `Greenland', cex=1.5, font=3)

#Add the locations of nine stations where bottom trawling was done
#These location variables may also be added from a data set:
lon <- c(-39.48,-32.97,-20.85,9.92,13.78,18.35, -5.55,3.76,12.03)
lat <- c(64.07, 66.84, 70.06,78.30,77.10,76.38,59.41,62.07,67.98)
points(lon, lat, col=`red', pch=19)
```

Example 2   *Create a map of Norway and plot the location of Masfjord.*
Use the *sp* library to create a map of Norway and show the location of Masfjord,
the fjord where the example data were collected.

Download the map of Norway from http://www.gadm.org. Download level 0, a
filecalled*NOR_adm0.rds*andusetheformatcalled"R(SpatialPolygonsDataFrame)";
this file is included in the example data. Save this file under `rwork/data`.

Remember to set the working directory of R to the `rwork` directory as
explained in Section 5.2

```
##########
# Norway #
##########

#Load the required library:
library(sp)

#Read in the rds-file containing the map as polygon data:
Norway <- readRDS(`data/NOR_adm0.rds')

#Plot the map:
plot(Norway, col=`lightgrey', border=`black')

#Add a coordinate point marking Masfjord on the map
points(5.446, 60.877, pch=19, cex=2, col=`red')

#Save map:
savePlot(`figures/Norway.png', type=`png')
```

The map should look similar to Figure 5.23.

Example 3   *Create a map of Masfjord.*
To create a map that shows only Masfjord, use the syntax above but specify
x- and y-axis limits using the `xlim` and `ylim` statements; these correspond to
longitude and latitude limits of the plot region. Add axis titles using the `xlab`
and `ylab` statements as shown in the example below. The `axes=T` statement is
necessary to plot the axes with the coordinate values. Set the color of the land
area by `col=...`, the border around land by `border=...`, and set the background
color (here, the sea) by `bg = ...` .

```
####################
# Masfjord, Norway #
####################
```

Figure 5.23 A map of Norway with the location of Masfjord shown as a red dot.

```
#Use coordinates that define Masfjord
#Add axes to show coordinate values (longitude and latitude):
plot(Norway,col='lightgrey',border='black',xlim=c(5.3,5.6), ylim=c
  (60.75,60.90),bg='lightblue',axes=T,xlab='Longitude',
  ylab='Latitude')
```

Add sampling stations to the plot using the same syntax to add points to a standard scatterplot (see Section 5.5.3). The station coordinates may be a variable in a data frame; here, they are vectors made specifically for this figure.

```
#Add 4 stations to the map - create the coordinates:
lon <- c(5.358785, 5.405737, 5.446566, 5.481678)
lat <- c(60.869588,60.871954,60.877139,60.882107)
stations <- c('1', '2', '3', '4')
text(cbind(lon, lat), labels=stations, cex=1, col='red')
```

Save the plot:

```
savePlot('figures/Masfjorden.png', type='png')
```

The map should look similar to the one in Figure 5.24. Save the plot:

```
savePlot('figures/Masfjorden.png', type='png')
```

The map in Figure 5.24 is very coarse resolution. This is because the downloaded base map does not contain a lot of detail. To create a map in higher resolution, find another map source that has high resolution spatial polygons or move to *Example 4.*

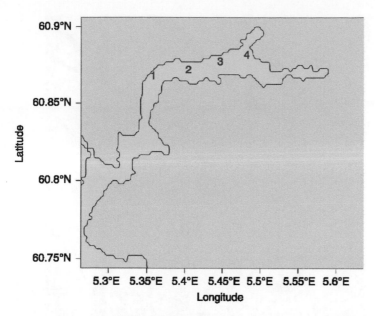

Figure 5.24 Low resolution map of Masfjord created with the *sp* package. The four sampling stations are marked as red numbers.

Example 4 *Create a high resolution map of Masfjord.*
The easiest way to improve the level of detail is to use *RgoogleMaps*, which can use maps from Google Maps.

```
######################################
# Masfjord, Norway, with topography #
######################################

#Load the library into the memory of R:
library(RgoogleMaps)

#Create a map object where longitude and latitude limits
#are specified by lonR and latR:
masfjord<-GetMap.bbox(lonR=c(5.3,5.6),latR=c(60.75,60.90),
  maptype='terrain')
```

This makes a terrain map object, but does not plot it. Replace "terrain" with "satellite" to create a map from a satellite picture.

```
#Plot the map by using the station positions created in previous
  example
PlotOnStaticMap(masfjord, lat=lat, lon=lon, cex=1,
  pch=stations,col='red')
```

Save the plot:

```
savePlot('figures/Masfjorden2.png', type='png')
```

The map should look similar to the one in Figure 5.25.

Figure 5.25 Map of Masfjord created with the *RgoogleMaps* package, which uses maps from Google Maps. The four sampling stations are marked as red numbers.

References

Becker, R.A., Wilks, A.R. and Brownrigg, R. (2016a) mapdata: Extra Map Databases. R package version 2.2-6. https://CRAN.R-project.org/package=mapdata (accessed June 23, 2017)

Becker, R.A. Wilks, A.R., Brownrigg, R. *et al.* (2016b) maps: Draw Geographical Maps, https://CRAN.R-project.org/package=maps (accessed June 23, 2017)

Bivand, R.S., Pebesma, E. and Gomez-Rubio, V. (2013) *Applied Spatial Data Analysis with R*, 2nd edn. Springer, New York, http://www.asdar-book.org/

Braga, R.R., Bornatowski, H. and Vitule, J.R.S. (2012) Feeding ecology of fishes: an overview of worldwide publications. *Reviews in Fish Biology and Fisheries*, 22, 915–929.

Bray, J.R. and Curtis J.T. (1957) An ordination of upland forest communities of southern Wisconsin. *Ecological Monographs*, 27, 325–349.

Kelley, D., Richards, C., Layton, C. *et al.* (2016) *oce: Analysis of Oceanographic Data.* https://CRAN.R-project.org/package=oce (accessed June 23, 2017)

Loecher, M. and Ropkins, K. (2015) RgoogleMaps and loa: unleashing R graphics power on map tiles. *Journal of Statistical Software*, 63, 1–18.

Oksanen, J., Blanchet, F.G., Friendly, M. *et al.* (2016) *vegan: Community Ecology Package*. https://CRAN.R-project.org/package=vegan (accessed June 23, 2017)

Pebesma, E.J. and Bivand, R.S. (2005) Classes and methods for spatial data in R. *R News* 5 (2), http://cran.r-project.org/doc/Rnews/ (accessed June 23, 2017)

Pinheiro, J.C. and Bates, D.M. (2000) *Mixed-effects models in S and S-PLUS*, Springer, New York.

R project core team (2016) *R: A Language and Environment for Statistical Computing*, https://www.R-project.org/ (accessed June 23, 2017)

Zuur, A., Ieno, E.N., Walker, N. *et al.* (2009) *Mixed Effects Models and Extensions in Ecology with R*, Springer, New York.

Index

Marine Ecological Field Methods: A Guide for Marine Biologists and Fisheries Scientists,
First Edition. Edited by Anne Gro Vea Salvanes, Jennifer Devine, Knut Helge Jensen,
Jon Thomassen Hestetun, Kjersti Sjøtun and Henrik Glenner.
© 2018 John Wiley & Sons Ltd. Published 2018 by John Wiley & Sons Ltd.

Printed and bound by CPI Group (UK) Ltd, Croydon, CR0 4YY

27/10/2024

14580304-0001